Contents

Preface

The public switched telephone network in the United States of America is one of the true marvels of the modern world. It provides the ability to interconnect any two out of more than one hundred million telephones, usually within a few seconds of the request for connection. It is controlled by the world's largest network of interconnected and cooperating computers. Yet the telephones in this network are usable by unskilled operators without formal training (almost any child of four or five can make a telephone call).

This book is in part about that network and the technology that has made it possible. But it is more about revolutionary changes that are taking place in the way telephone conversations and data are taken into the network, switched, and transmitted. Some of the technology still in everyday use in the telephone network dates from the decade of the invention of the telephone in the 1870s. However, even this old and traditional business is being forced by economics, regulation, and competition to make massive changes in the way it does business and in the equipment and techniques used to provide the telephone service.

We begin by presenting the fundamentals of the telephone network: how it began, what the components are, and how they are connected together. Next we review the basic nonelectronic telephone set. We then consider the effect of microelectronics on the construction and operation of the telephone set; for example, the effect on functions such as speech signal processing and interface with the telephone line, pulse and tone generation for dialing, and ringers. Next, we discuss how microcomputers use digital techniques and stored programs to enhance the performance and features of the telephone set. Digital transmission techniques, electronics in the central office, and network transmission concepts and fundamentals are explained. Finally, we briefly consider the relatively new cordless phone set and the new concept of mobile radio and cellular phone service.

Like other books in the series, this book builds understanding step-by-step. Try to master each chapter before going on to the next one. A quiz is provided at the end of each chapter for self-evaluation of what has been learned. Answers are given at the back of the book.

The business of providing the equipment and service for both local and long distance telephone communication is today undergoing some of the most fundamental changes ever required to be made by any U.S. institution. Before the government breakup of AT&T in January 1984, it was the largest company on earth, and its product is a service that has become a necessity for all of us. Understanding the technical side of the telephone is necessary in understanding a force in modern life which has been and will continue to be as much an agent for change as the automobile, the airplane, and the computer.

Acknowledgments

First Edition:

The authors wish to acknowledge the tutelage of Dr. John Bellamy, whose knowledge of this subject fills his students and this book. They are grateful also for the friendship and advice of John McNamara of Digital Equipment Corporation, Ken Bean of Texas Instruments, and Leo Goeller, who always has time to chat. Finally, we are grateful to the members of the International Communications Association, whose encouragement and support have changed our lives.

Figures reprinted with permission of Bell Telephone Laboratories and AT&T came from the list that follows. Other publications are credited individually.

1. Members of the Technical Staff, Bell Telephone Laboratories, *Engineering and Operations of the Bell System*, Bell Telephone Laboratories, Inc., 1977.

2. Members of the Technical Staff, Bell Telephone Laboratories, *A History of Engineering and Science in the Bell System, The Early Years (1875–1925)*, Bell Telephone Laboratories, Inc., 1975.

3. Members of the Technical Staff, Bell Telephone Laboratories, *Bell Laboratories Record*, November, 1980.

4. Network Planning Division of AT&T, *Notes on the Network*, American Telephone and Telegraph Co., No. 500-029, 1980.

Third Edition:

I would like to thank Mr. Sheldon Hochheiser of the AT&T Archives for providing ample information on the history and break-up of AT&T. I would also like to acknowledge the following individuals for their generous permission to reprint material:

1. Mr. T. M. Dalton III, Manager, Business Services, Texas Instruments, Inc.

2. Mr. Mark B. Jorgensen, Director, Corporate Communications, Silicon Systems, Inc.

3. Mr. L. Jefferson Gorin, Manager, Media Relations—Phoenix, Motorola Semiconductor Products—Phoenix, Arizona.

4. Ms. Eileen Algaze, Public Relations Manager, Marketing Communications, Rockwell International Corporation.

The Telephone System

INTRODUCTION

The telephone was invented a little over a hundred years ago by Alexander Graham Bell. The telephone industry has since become one of the largest on earth.

The telephone arrived as a practical instrument over a century ago in 1876, an outgrowth of experiments on a device to send multiple telegraph signals over a single wire. Alexander Graham Bell, a native of Scotland, while conducting electrical experiments spilled acid on his trousers. His sulphurous reaction, the now famous "Mr. Watson, come here, I want you," brought Thomas A. Watson on the run not only because of his employer's distress, but because the words had been carried by electricity into Watson's room and reproduced clearly on his receiving set. The simple instrument being tested on Court Street in Boston on March 10, 1876 wasn't very practical (the acid was used in the system), but improvement followed so rapidly that putting into action Bell's concept of a public telephone network—"this grand system . . . whereby a man in one part of the country may communicate by word of mouth with another in a distant place"—was well underway by January of 1878, when the first commercial exchange was operated in New Haven. By 1907, one hotel alone (the Waldorf Astoria in New York City) had 1,120 telephones and processed 500,000 calls per year.

The American Telephone and Telegraph Company (AT&T) was incorporated in March of 1885 to manage the explosive growth of the fledgling telephone network across the United States. Virtually since its beginning, AT&T worked as a legal, regulated monopoly. This means that AT&T was allowed to establish, maintain, and control a single, universal network across the country without any competition, as well as provide all telephone sets and switching equipment to the general public. The federal government regulated its policies, practices, and fees. This set the groundwork for the development of the most advanced and efficient telecommunications system in the world.

By the mid-1940s however, the U.S. government began to question seriously the principles of the telephone monopoly in light of the general antitrust laws and alleged abuses by AT&T. An antitrust suit filed in 1949 forced AT&T to restrict its business activities to the national telephone system in 1956. Over the next several decades, the Federal Communications Commission (FCC) began to allow the introduction of new products and services from competing companies. By the mid-1970s, several competitors obtained the capacity to offer long-distance telephone service.

Advances in technology and the challenges of competition caused the government once again to rethink its position on the telephone monopoly. On November 20, 1974, the Department of Justice filed a new antitrust suit against AT&T. The trial began in January of 1981. One year later, AT&T agreed on terms to settle the suit. In essence, AT&T would divest all of its local operating companies. This would dissolve the monopoly held by AT&T for almost 100 years, but it would also lift many regulatory constraints.

Official divestiture took place on January 1, 1984. The monopoly was gone. AT&T was free to compete in the nation's emerging communications market, and local operating companies were allowed to handle local service and maintain the network. *Table 1-1* shows the regional distribution of local telephone companies.

**Table 1-1.
Regional Telephone
Companies After
Divestiture**

Region	Local Companies
Northeast Region	New England Telephone New York Telephone
Mid-Atlantic Region	Bell of Pennsylvania The Chesapeake and Potomac Telephone Companies New Jersey Bell
Southeast Region	South Central Bell Southern Bell
Midwest Region	Illinois Bell Indiana Bell Michigan Bell Ohio Bell Wisconsin Bell
Southwest Region	Southwestern Bell
US West Region	Mountain Bell Northwestern Bell Pacific Northwest Bell Pacific Telephone

Each local telephone company is owned by a holding company for that particular region which handles the overall day-to-day business operations of the region, and leaves the local companies to concentrate on service and maintenance of the network. These holding companies are known as Regional Bell Operating Companies (RBOCs). In order to maintain technical consistency after divestiture, a Central Service Organization (CSO) was established to serve RBOCs across the nation by providing research and development functions. This technical organization takes the place of Bell Labs which remained with AT&T after divestiture. The CSO is funded by every RBOC, so its work is utilized by all regional and local companies.

Today, telecommunications is a multibillion dollar industry employing well over one million people. This modern network handles voice and data communications efficiently and reliably in even the most remote locations.

THE TELEPHONE SET

The telephone set performs 8 electrical functions to provide us with service.

Telephone sets like those used to originate and receive telephone calls are shown in *Figure 1-1*. It is simple in appearance and operation yet it performs a surprising number of functions. The most important ones are:

1. It requests the use of the telephone system when the handset is lifted.
2. It indicates that the system is ready for use by receiving a tone, called the dial tone.
3. It sends the number of the telephone to be called to the system. This number is initiated by the caller when the number is pressed or the dial is rotated.
4. It indicates the state of a call in progress by receiving tones indicating the status (ringing, busy, etc.).
5. It indicates an incoming call to the called telephone by ringing bells or other audible tones.
6. It changes speech of a calling party to electrical signals for transmission to a distant party through the system. It changes electrical signals received from a distant party to speech for the called party.
7. It automatically adjusts for changes in the power supplied to it.
8. It signals the system that a call is finished when a caller "hangs-up" the handset.

Of course, for a telephone to be of any use, it must be connected to another telephone. In the very early days of telephony, the phones were simply wired together with no switching. As the number of phones increased this became impractical, so the local exchange or central office was established to handle the switching and other functions.

THE LOCAL LOOP

A single pair of wires connects the telephone to the central switching office. This connection is called a local loop. One connection is called the tip (T) and the other connection the ring (R).

Each subscriber telephone is connected to a central office that contains switching equipment, signaling equipment, and batteries that supply direct current to operate the telephone as shown in *Figure 1-2*. Each phone is connected to the central office through a local loop of two wires called a wire pair. One of the wires is called T (for tip) and the other is called R (for ring) which refers to the tip and ring parts of the plug used in the early manual switchboards.

Figure 1-1.
Telephone Set
(Courtesy Radio Shack)

a. Rotary Dial

b. Keypad Dial

**Figure 1-2.
Telephone Set and
Central Office
Exchange Simplified
Circuits** *(Source: D. L.
Cannon and G. Luecke,*
Understanding
Communications
Systems, *SAMS, a
Division of Macmillan
Computer Publishing,
1984)*

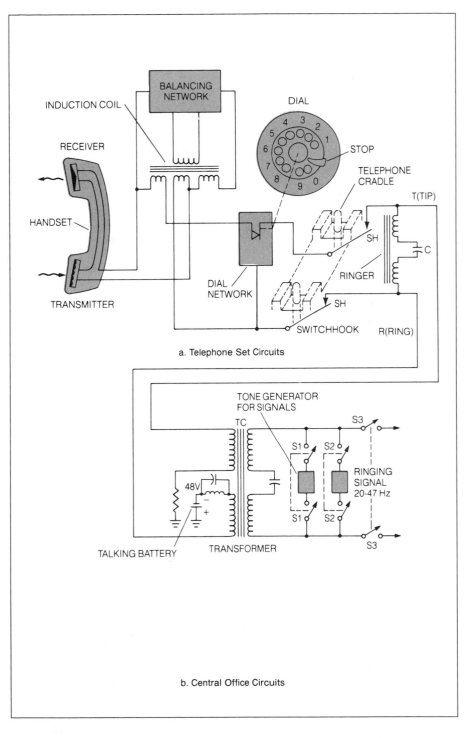

a. Telephone Set Circuits

b. Central Office Circuits

Switches in the central office respond to the dial pulses or tones from the telephone to connect the calling phone to the called phone. When the connection is established, the two telephones communicate over transformer coupled loops using the current supplied by the central office batteries.

Initiating a Call

When the "receiver" handset is in the off-hook condition, the exchange sends a dial tone to the calling telephone.

When the handset of the telephone is resting in its cradle, the weight of the handset holds the switchhook buttons down and the switches are open. This is called the on-hook condition. The circuit between the telephone handset and the central office is open; however, the ringer circuit in the telephone is always connected to the central office as shown in *Figure 1-2*. The capacitor, C, blocks the flow of dc from the battery, but passes the ac ringing signal. (The ringer circuit presents a high impedance to speech signals so it has no effect on them.)

When the handset is removed from its cradle, the spring-loaded buttons come up and the switchhook closes. This completes the circuit to the exchange and current flows in the circuit. This is called the off-hook condition. (The on-hook, off-hook, and hang-up terms came from the early days of telephony, when the receiver was separate and hung on the switchhook when not in use as shown in *Figure 1-3*. This also explains why many people still refer to the handset of today as the receiver.)

The off-hook signal tells the exchange that someone wants to make a call. The exchange returns a dial tone to the called phone to let the caller know that the exchange is ready to accept a telephone number. (The telephone number also may be referred to as an address.)

Sending a Number

Some telephone sets send the telephone number by dial pulses while others send it by audio tones.

Dial Pulsing

Numbers are sent either by a stream of pulses (pulse dialing) or by a series of audio tones (tone dialing).

Telephone sets that use dial pulsing have a rotary dial as shown in *Figure 1-1a,* which opens and closes the local loop circuit at a timed rate. The number of dial pulses resulting from one operation of the dial is determined by how far the dial is rotated before releasing it.

Although all network facilities are currently compatible with pulse dialing telephones, today's standard embraces the tone method of dialing.

Dual Tone Multifrequency (DTMF)

Most modern telephone sets employ the newer method of using audio tones to send the telephone number. These can be used only if the central office is equipped to process the tones. Instead of a rotary dial, these telephone sets have a push-button keypad with 12 keys for the numbers 0 through 9 and the symbols * (asterisk) and # (pound sign).

**Figure 1-3.
Early Telephone with
Separate Receiver
Hanging on
Switchhook.**

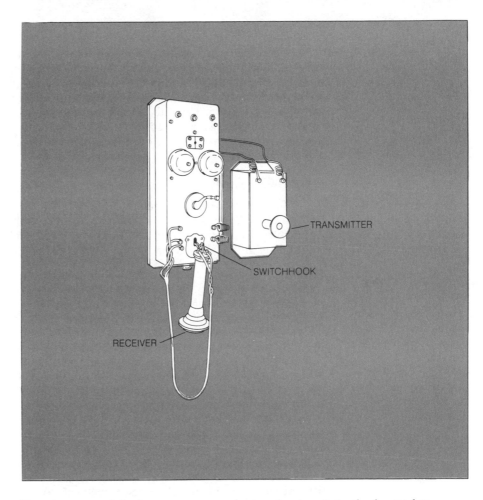

TRANSMITTER

SWITCHHOOK

RECEIVER

Pressing one of the keys causes an electronic circuit in the keypad to generate two output tones that represent the number.

Connecting the Phones

The connection having been made at the switching office, a ringing signal is sent to the called telephone.

The central office has various switches and relays that automatically connect the calling and called phones. For now, assume that the connection has been made. The actual operation of switching systems will be covered in more detail a little later.

If the called phone handset is off-hook when the connection is attempted, a busy tone generated by the central office is returned to the calling phone. Otherwise, a ringing signal is sent to the called phone to alert the called party that a call is waiting. At the same time, a ringback tone is returned to the calling phone to indicate that the called phone is ringing.

Ringing the Called Phone

Early telephone circuits were point-to-point (not switched), and the caller gained the attention of the party at the other end by picking up the transmitter and shouting "Hello" or "Ahoy." This was not very satisfactory, and schemes based on a mechanical signaling arrangements were soon invented. The one in common use today, called the "polarized ringer," or bell, was patented in 1878 by Thomas A. Watson (Mr. Bell's assistant). Electronic ringing circuits are quickly replacing polarized ringers in new telephone designs.

Answering the Call

Removing the handset at the ringing telephone results in a loop current flow.

When the called party removes the handset in response to a ring, the loop to that phone is completed by its closed switchhook and loop current flows through the called telephone. The central office then removes the ringing signal and the ringback tone from the circuit.

Talking

The transmitter converts acoustical energy into equivalent electric current variations. The receiver converts these electrical variations into the equivalent acoustical energy— called sound.

The part of the telephone into which a person talks is called the transmitter. It converts speech (acoustical energy) into variations in an electric current (electrical energy) by varying or modulating the loop current in accordance with the speech of the talker.

The part of the telephone that converts the electric current variations into sound that a person can hear is called the receiver. The signal produced by the transmitter is carried by the loop current variations to the receiver of the called party. Also, a small amount of the transmitter signal is fed back into the talker's receiver. This is called the sidetone.

Sidetone is necessary so that the person can hear his/her own voice from the receiver to determine how loudly to speak. The sidetone must be at the proper level because too much sidetone will cause the person to speak too softly for good reception by the called party. Conversely, too little sidetone will cause the person to speak so loudly that it may sound like a yell at the receiving end.

Ending the Call

If either telephone handset is hung up, the current loop is opened and the central office releases the line connection.

The call is ended when either party hangs up the handset. The on-hook signal tells the central office to release the line connections. In some central offices, the connection is released when either party goes on-hook. In others, the connection is released only when the calling party goes on-hook.

Beyond the Local Loop

Thus far, the discussion of connecting two telephones together has been limited to local loops and a central office exchange. Most central office exchanges can handle up to 10,000 telephones. But what if it is required to connect more than 10,000 phones, or to connect phones in different cities, different states, or different countries? Over the years, a complex network of many telephone exchanges has been established to accomplish these requirements. Let's look next at how this network is arranged.

THE PUBLIC SWITCHED TELEPHONE NETWORK

Exchange Designations

Telephone exchanges exist in a network hierarchy. Usually the first four classes are for long distance switching, and the fifth for connection to the subscription telephones.

Each telephone exchange in North America has two designations, office class and name, to identify it and to describe its function. These are shown in *Figure 1-4.*

Subscriber telephones are normally, but not exclusively, connected to End Offices. Toll (long distance) switching is performed by Class 4, 3, 2, and 1 offices. The Intermediate Point, or Class 4X office, is a relatively new class. It applies to all-digital exchanges to which remote unattended exchanges (called Remote Switching Units) can be attached. These Class 4X offices may interconnect subscriber telephones as well as other Class 5 and Class 4 exchanges.

The ten Regional Centers (Class 1 offices) in the U.S. and two in Canada are connected directly to each other with large-capacity trunk groups. In 1981, there were 67 Class 2; 230 Class 3; 1,300 Class 4 and about 19,000 Class 5 exchanges in North America.

Interconnection

The network attempts to make connection at the lowest possible level, and therefore the shortest path. If the lines are all busy, trunk groups at the next highest level are used.

The network is organized like a tree, or rather like a small grove of trees, whose roots have grown together. *Figure 1-4* shows this in simplified form. Each exchange is optimized for a particular function. A call requiring service which cannot be performed by a lower class exchange is usually forwarded to the next higher exchange in the network for further processing.

The Regional Center, like the base of each tree, forms the foundation of the network. The branch levels are the Class 2, 3, 4, 4X, and Class 5 offices. Most offices are connected to more than one other, and the interconnections among the various offices are not as simple as shown in *Figure 1-4.* The interconnections depend on the patterns of the traffic arriving at and leaving each office.

The network makes connections by attempting to find the shortest path from the Class 5 office serving the caller to the Class 5 office serving the called party. The high-usage interoffice trunk groups which provide direct connection between offices of equal or lower level are used first. If they are busy, trunk groups at the next higher level (called final groups) are used. Digital logic circuits in the common control of each exchange make decisions based on rules stored in memory that specify which trunk groups are to be tried and in what order. These rules, for example, prevent more than nine connections in tandem, and prevent endless loop connections (called ring-around-the-rosy).[1]

[1]Nine tandem connections have never been known to occur.

Figure 1-4.
Network Hierarchy

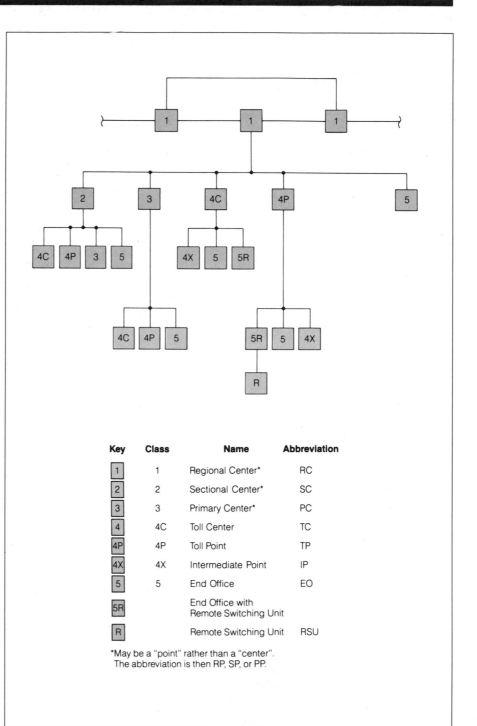

Structure

The control and voice signals are carried by three types of facilities—local, exchange area, and long-haul.

The supervisory signals used to set up telephone connections and the voice signals of the conversations are carried by transmission systems over paths called facilities. These systems are divided into three broad categories—Local, Exchange Area, and Long-Haul.

The Local Network

The local network consists of homes and businesses connected via wire pairs to a central office.

The local network shown in *Figure 1-5* is the means by which telephones in residences and businesses are connected to central offices. The local facilities are almost exclusively wire pairs which fan out like branches of a tree from a point called the wire center throughout a serving area. Serving areas vary greatly in size, from an average of 12 square miles in urban locations to 130 square miles for rural areas. More than one central office is often required for a serving area in urban areas, but one central office is usually sufficient in rural areas. An average wire center in an urban area will serve 41,000 subscriber lines and 5,000 trunks. The urban exchanges are generally of higher call carrying capacity than the rural exchanges.

The Exchange Area Network

The exchange area network fills the transmission gap between local and long-distance trunks.

The exchange area network is intermediate between the local network and the long-haul network. A simplified example is shown in *Figure 1-6*. Exchanges are interconnected with exchange area transmission systems. These systems may consist of open wire pairs on poles, wire pairs in cables, microwave radio links, and fiber optic cables. The exchange area network normally interconnects local exchanges and tandem exchanges. Tandem exchanges are those that make connections between central offices when an interoffice trunk is not available. A tandem exchange is to central offices as a central office is to subscriber telephone sets.

The Long-Haul Network

The long-haul network of the 3 major US carriers is made up primarily of high capacity fiber optic cables.

During the 1980s, wire cable and microwave transmission facilities connecting local and toll (long-distance) exchanges in the United States were aggressively replaced with fiber optic cables. Of the three major long-distance carriers, U.S. Sprint's network is 100% fiber optic. The networks operated by AT&T and MCI are about 75% fiber optic. It is projected that 100% of U.S. long-distance transmission media will be made up of optical fiber by the mid-1990s. Satellite and microwave links will be used in situations where optical fiber installation is not practical or economical.

Figure 1-5.
Local Network
(Courtesy Bell
Laboratories)

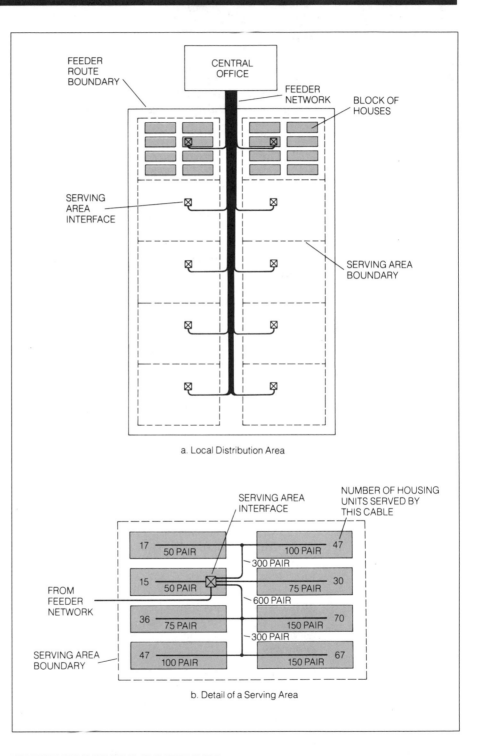

a. Local Distribution Area

b. Detail of a Serving Area

**Figure 1-6.
Exchange Area
Network** *(Courtesy Bell
Laboratories)*

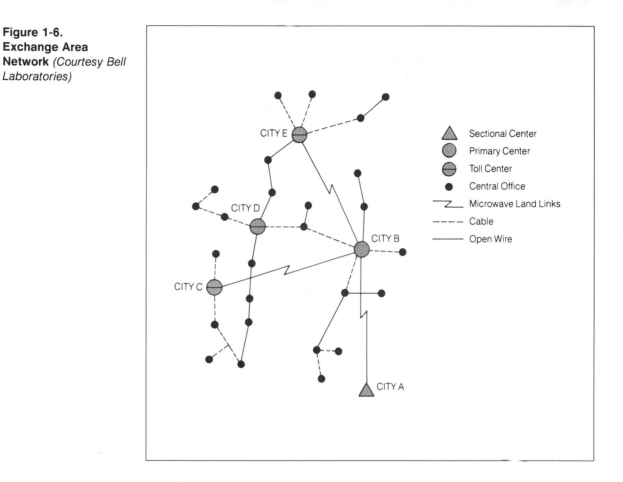

TYPES OF TRANSMISSIONS

Both voice and control signals are carried. Voice signals are usually analog, but control signals or supervisory signals may be digital and/or analog.

Spoken messages or voice signals are not the only signals that are transmitted down a telephone line. In the previous discussion of making a connection between the calling telephone and the called telephone, some of these other signals were discussed: dial tone, dial pulses or key tones used for sending a number, busy tone, and ringback tone. These signals are for control of the switching connections or to indicate the status of the call. Such signals are called control signals or supervisory signals. They may be tone signals (analog) or On-Off (digital) signals. Therefore, if one were to examine the signals on many local loops, one would find analog voice signals, analog tone signaling, and digital On-Off signaling. It would be a mixture of analog and digital signals.

Analog Voice Transmissions

Signals that have continuously and smoothly varying amplitude or frequency are called analog signals. Speech (or voice) signals are of this type. They vary in amplitude and frequency. *Figure 1-7* shows the typical distribution of energy in voice signals. The vertical axis is relative energy and the horizontal axis is frequency. It shows that the voice frequencies that contribute to speech can extend from below 100 hertz (Hz) to above 6,000 Hz. However, it has been found that the major energy necessary for intelligible speech is contained in a band of frequencies between 200 Hz and 4,000 Hz.

Figure 1-7.
Voice Energy
Frequency

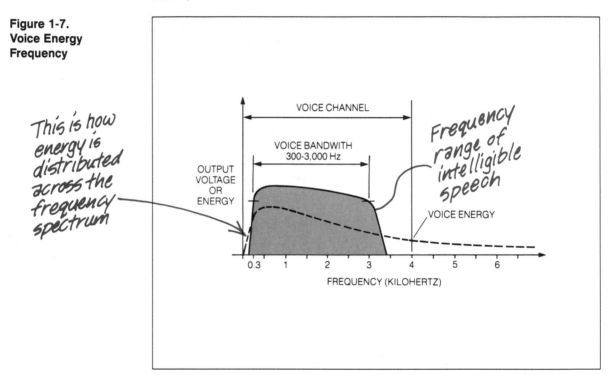

This is how energy is distributed across the frequency spectrum

Frequency range of intelligible speech

Voice Channel Bandwidth

The telephone circuits are designed to pass a limited bandwidth. This permits the transmission of the voice frequencies and limits unwanted circuit noises.

In order to eliminate unwanted signals (noise) that could disturb conversations or cause errors in control signals, the circuits that carry the telephone signals are designed to pass only certain frequencies. The range of frequencies that are passed are said to be in the pass band. For a telephone system voice channel (a VF channel) the pass band is 0 to 4,000 Hz. (Sometimes this band is called a message channel.) The bandwidth is the difference between the upper and lower limits of the pass band; thus, the bandwidth of the VF channel is 4,000 Hz. However, not all of the VF channel is used for the transmission of speech. The voice pass band is restricted to 300 to 3,000 Hz as shown in *Figure 1-7*. Hence, any signal carried on the telephone circuit which is within the range of 300 to 3,000 Hz is called an in-band signal as shown in *Figure 1-8.* Any signal which is not within the 300 to 3,000 Hz band, but is within the VF channel, is called an out-of-band signal. All speech signals are in-band signals. Some signaling transmissions are in-band and some are out-of-band.

Figure 1-8.
In-Band and Out-of-Band Signaling

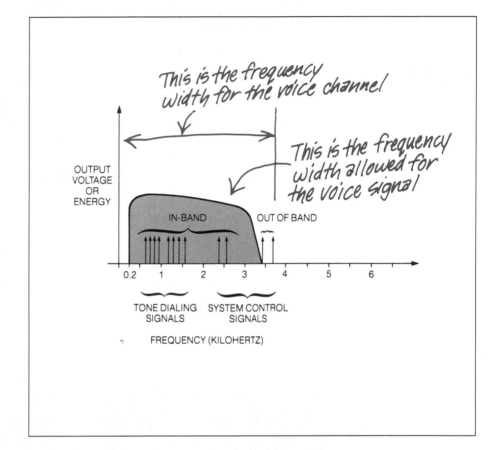

Voice Channel Level

The loudness or amplitude of signals on telephone circuits is usually referred to as the level of the signal. The level of a signal is expressed in terms of the power which the signal delivers to a load. For example, a pair of telephone wires running together in a cable forms a transmission line with an impedance of 600 ohms. Impedance is to ac circuits what resistance is to dc circuits. As shown in *Figure 1-9,* the power delivered to a balanced pair transmission line is:

$$P_{load} = \frac{e_s^2}{Z}$$

where,
      P_{load} is the power in watts,
      e_s is the signal level in volts,
      Z is the impedance in ohms.

**Figure 1-9.
Power Delivered to
Wire Pair**

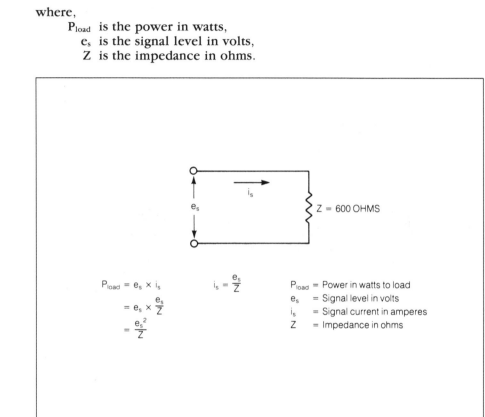

Signal level usually is expressed relative to some reference. In telephone and audio circuits, the reference level is 1 milliwatt of power to the load. If P_{load} equals 1 milliwatt (0.001 watt) and Z equals 600 ohms then, as shown in *Figure 1-9:*

$$1 \; mW = \frac{e_s^2}{600 \; ohms}$$

or

$$600 \times 1 \times 10^{-3} = e_s^2$$
$$0.6 = e_s^2$$
$$0.775 = e_s$$

The signal level at any point in the circuit is referenced to a power level of one milliwatt, and referred to as 0 dBm.

Therefore, a signal level of 0.775 volt applied across 600 ohms produces 1 milliwatt of power.

Analog signals that are transmitted at a constant frequency also can have their level expressed in decibels (dB). It is another means of expressing the signal power delivered to a load. In technical terms as an equation:

dB $= 10 \log_{10} (P_1/P_2)$

It is a shorthand way of expressing the ratio of power P_1 to power P_2. *Table 1-2* lists some of these ratios.

A special decibel ratio is established when 1 milliwatt is used as the reference power, P_2. Under this condition, the dB power ratio is classified as measured in dBm (*decibels* referenced to 1 *m*illiwatt). Therefore, from *Table 1-2,* if P_2 equals 1 milliwatt, then a signal at 0 dBm will be delivering a power, P_1, of 1 milliwatt to the load because the ratio of P_1 to P_2 must be 1. Said another way, when a signal produces a power P_1 into a load of 600 ohms that has a 20 dBm level, it is delivering 100 milliwatts of power (P_1) compared to the reference power of 1 milliwatt for P_2.

Table 1-2.
Power Ratios in dB

dB	P1/P2
40	10,000
30	1,000
20	100
10	10
3	2.0
0	1.0
-3	0.5
-10	0.1
-20	0.01
-30	0.001
-40	0.0001

In telephone systems, the 0 dBm level is usually set at the sending end of a transmission line at the output of the switch. This point then becomes a system reference point called the zero transmission level point (0 TLP). Once the 0 TLP is chosen and the 0 dBm level applied at that point, all other power gains and losses in the transmission path between that point and the next switch output can be measured directly with respect to the 0 TLP. If the signal magnitude is measured, then the unit dBm0 is used. If only the relative gain or loss is indicated, the unit dB is used.

Voice Channel Noise

Unwanted signals in the voice frequencies are called noise. Any source of electrical energy has the potential for inducing noise on the lines.

Transmission systems often must operate in the presence of various unwanted signals (referred to generally as noise) that distort the information being sent. Lightning, thermal noise, induced signals from nearby power lines, battery noise, corroded connections, and maintenance activities all contribute to degradation of the signal. Analog channel speech quality is primarily determined by the absolute noise level on the channel when it is idle; that is, when no speech signal is present. Speech tends to mask any noise present, but noise in an idle channel is quite objectionable to a listener. Stringent standards (−69 dBm0 up to 180 miles and −50 dBm0 up to 3,000 miles with −16 dBm0 as speech level) have been set for this idle channel noise in the U.S. network.

Echoes are caused by the reflection of unabsorbed electrical energy by the load side of the transmission lines.

Another type of noise that originates from the voice transmission itself is an echo. The primary echo is the reflection of the transmitted signal back to the receiver of the person talking. The amount of delay in the echo depends on the distance from the transmitter to the point of reflection. The effect of the delay on the talker may be barely noticeable to very irritating to downright confusing. Echo also affects the listener on the far end, but to a lesser degree. Echoes are caused by mismatches in transmission line impedances which usually occur at the hybrid interface between a 2-wire circuit and a 4-wire transmission system. The effect of echo is reduced by inserting loss in the lines.

Multiplexing

Multiple telephone conversations may be sent over one telephone channel by frequency division multiplexing (FDM). This involves assigning each voice signal to a separate carrier frequency.

A local loop can carry only one voice channel conversation at a time. This is not economical for toll transmission and a method was devised so that a transmission path can carry many telephone conversations at the same time. This is accomplished by multiplexing. For analog signals, frequency division multiplexing (FDM) is used. In simplified terms, this means that several telephone conversations are all sent together over one transmission channel, but are separated by their frequency.

The basic principles of this are shown in *Figures 1-10* and *1-11*. In *Figure 1-10,* a voice signal having frequencies within the voice frequency channel bandwidth of from 0 to 4 kilohertz (kHz) is changing, or modulating, the amplitude of another frequency (8,140 kHz in this case) which is called the carrier frequency. The 0–4 kHz voice frequency signal is amplitude modulating the 8,140 kHz carrier. The information in the voice signal is being carried by the changing amplitude of the 8,140 kHz signal and the voice frequencies have been translated to different frequencies. This technique is commonly referred to as amplitude modulation.

**Figure 1-10.
Carrier Modulated by
Voice Signal**

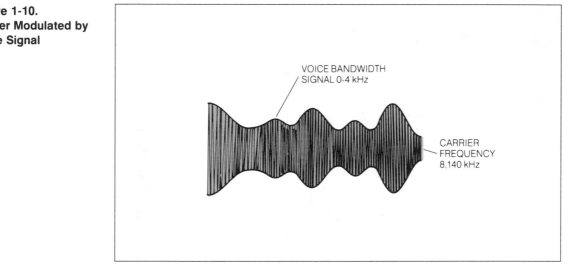

If different voice signals (different telephone conversations) are placed on different carrier frequencies, then many conversations may be multiplexed on one transmission path and transmitted to the receiving point. At the receiving point, the different conversations can be identified and separated by their unique frequency, and the original conversation can be recovered from the carrier (demodulated) and sent to the called telephone.

The multiplexing of the signals is shown in *Figure 1-11* for 12 voice channels. Since each voice channel has a 4 kHz bandwidth, 12 channels require 4×12 or 48 kHz bandwidth. Since the lower frequency in this example is 8,140 kHz, the output multiplexed signal frequency extends from 8,140 kHz to 8,188 kHz $(8,140 + 48 = 8,188)$. It should be apparent that if the individual voice channel bandpass were made larger, the spread in carrier frequency would have to be larger; or if the number of voice channels to be multiplexed together were increased, the spread in carrier frequencies would need to be larger. In technical terms, in general, as the number of voice channels to be transmitted over a transmission path increases, the required bandwidth of the transmission path must increase.

Signaling Transmission

As stated previously, signaling refers to specific signals on the transmission line that are used for controlling the connection from the calling telephone to the called telephone, or that are used to indicate the status of a call as it is being interconnected. The first type to be discussed is dc signaling.

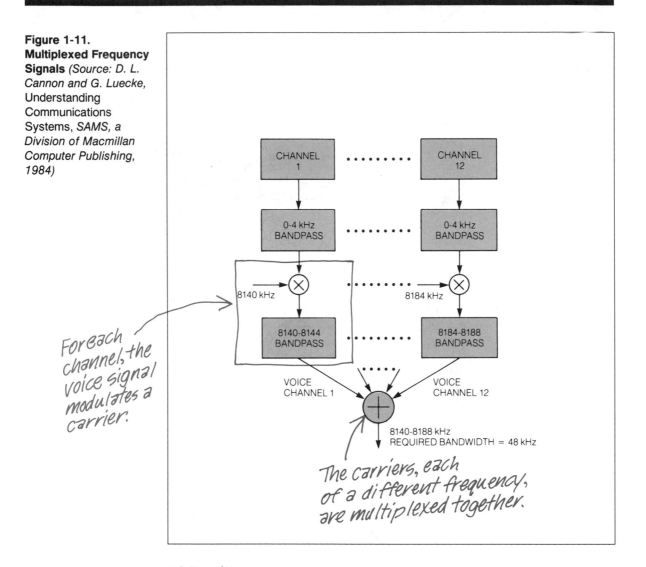

Figure 1-11.
Multiplexed Frequency
Signals *(Source: D. L.*
Cannon and G. Luecke,
Understanding
Communications
Systems, *SAMS, a*
Division of Macmillan
Computer Publishing,
1984)

DC Signaling

Telephone signaling
techniques and status
information transmittal
are performed by the
pulsing or polarity re-
versal of the loop line
current.

Dc signaling is based on the presence or absence of circuit current or voltage, or the presence of a given voltage polarity. The state of the signal indicates on-hook, off-hook, dial pulses, or status of the interconnection. These signals are On-Off type digital signals.

On local loops, on-hook is indicated by an open circuit and no current flow. Off-hook is signaled by a closed circuit and continuous current flow. Dial pulses consist of current flow interrupted at a specified rate as discussed previously. (A potential problem with dc signaling is that dial pulses spaced too far apart may be mistaken for an on-hook signal by the exchange. However, due to careful design, this problem does not occur very often.)

A type of dc signaling called reverse battery signaling is used between central offices to indicate the status of the switched connection. When the near end exchange requests service, an idle trunk is seized. A polarity of a given voltage exists on the trunk which indicates to the near end that the called phone is on-hook and ringing. The far end exchange acknowledges and indicates to the near end that the called party has answered by reversing the voltage polarity.

E & M signaling is used for the same purpose on long interoffice and short-haul toll trunks. This type signaling requires two extra wires in the originating and terminating trunk circuits—one for the E lead and the other for the M lead. Since separate wires are used for each, the on-hook and off-hook states can be signaled from both ends of the circuit as shown in *Table 1-3*. This allows signaling to be sent in both directions at the same time without interfering with one another. Sometimes two wires are used for each signal to avoid noise problems caused by a common ground.

**Table 1-3.
E & M Signaling**

State	E-Lead (Inbound)	M-Lead (Outbound)
On-hook	Open	Ground
Off-hook	Ground	Battery Voltage

Tone Signaling

Tone or analog signals indicate the status of the called telephone to the initiating caller and for control signals.

Various tones are used for both control and status indication. The tones may be single frequency or combinations of frequencies. These are analog signals that are either continuous tones or tone bursts (tones turned on and off at various rates). The call progress tones listed in *Table 1-4* are sent by the exchange to the calling phone to inform the caller about the status of the call. For example, the dial tone, which has been mentioned previously, is a continuous tone made by combining the frequencies of 350 Hz and 440 Hz. The busy signal that tells the caller that the called telephone is busy (off-hook) is a combination frequency tone that appears in bursts of 0.5 second on time separated by an off time of 0.5 second. The receiver off-hook warning signal is a combination frequency tone of four frequencies which is on for 0.1 second and off for 0.1 second. This signal is very loud in order to get the attention of someone to "hang up" the receiver (handset) that has been left off-hook. All of these tones, as well as the DTMF addressing tones discussed previously, are in-band signaling.

Tone signaling between exchanges may be in-band or out-of-band. The most commonly used single frequency (SF) tones are 2,600 Hz for in-band and 3,700 Hz for out-of-band signaling. E & M signals are converted to an SF tone for transmission on carrier systems because the dc signals cannot be transmitted. The tone indicates on-hook when present and off-hook when not present. Multifrequency (MF) signaling uses six

**Table 1-4.
Network Call Progress
Tones**

Tone	Frequency (Hz)	On Time (Sec)	Off Time (Sec)
Dial	350 + 440	Continuous	
Busy	480 + 620	0.5	0.5
Ringback, Normal	440 + 480	2	4
Ringback, PBX	440 + 480	1	3
Congestion (Toll)	480 + 620	0.2	0.3
Reorder (Local)	480 + 620	0.3	0.2
Receiver Off-hook*	1400 + 2060 + 2450 + 2600	0.1	0.1
No Such Number	200 to 400	Continuous, Frequency modulated at 1 Hz Rate	

Receiver off-hook is a very loud tone, 0 dBm per frequency

frequencies: 700; 900; 1,100; 1,300; 1,500; and 1,700 Hz for transmitting address information (the telephone number) over toll facilities. The frequencies are used in pairs to represent the numerals 0 through 9 and some control functions much like DTMF is used at the telephone set.

Control Signals That Are Digital

Control signals may also be digital codes. Each piece of telephone status or command information appears on the line as a unique combination of "highs" and "lows," 8-bits long.

Instead of just interrupting a dc voltage, as in the case of dc signaling, or interrupting continuous tones to provide tone bursts, control or supervisory signals also can be digital codes. Instead of being On-Off signals that occur at random times, they are combinations of signals that have two levels, 0 and 1, and that have a definite time relationship with each other. This is illustrated in *Figure 1-12*. In the telephone system the binary digit (bit) 1 and 0 levels shown may be represented by voltage or current levels. Note that the bits occur in a particular time sequence. For example, in *Figure 1-12,* a binary code of 8 bits is shown with bits d_0 through d_7 always occurring in the same time slot, t_1 through t_8, when transmitted in sequence. For a particular system design, once the time relationships of when the bits occur is set, the time relationship doesn't change.

The control information can be contained in the binary code in several ways. All 8 bits may be used as a group to represent a number from 0 to 255. The binary code for the number 234 is shown across the first line in *Figure 1-12*. On the left side, the code is presented in 1s and 0s, and on the right side, the code is presented as voltage levels or pulses. Or the 8-bit group may represent a letter of the alphabet in a data

Figure 1-12.
Serial Digital Code
Signals

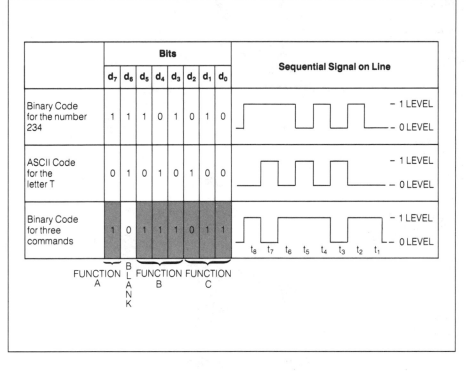

	Bits								Sequential Signal on Line
	d_7	d_6	d_5	d_4	d_3	d_2	d_1	d_0	
Binary Code for the number 234	1	1	1	0	1	0	1	0	– 1 LEVEL – 0 LEVEL
ASCII Code for the letter T	0	1	0	1	0	1	0	0	– 1 LEVEL – 0 LEVEL
Binary Code for three commands	1	0	1	1	1	0	1	1	– 1 LEVEL t_8 t_7 t_6 t_5 t_4 t_3 t_2 t_1 – 0 LEVEL

FUNCTION A · BLANK · FUNCTION B · FUNCTION C

communications code. A letter T in ASCII is shown on the second line. Or individual bits or subgroups of the 8-bit code may be used to command different functions. Examples of subgroup codes for the functions A, B, and C are shown on the third line.

Common Channel Interoffice Signaling

A special method of signaling (common channel interoffice signaling) uses no-voice circuits for control and addressing purposes. Computers manage the system.

All the signaling methods discussed so far send the control and addressing signals over the same circuit as the voice signals. Another method that is used separates the control signals from the voice signals. The control signals are sent over a separate circuit where they are detected and do the control and switching of lines independently from the voice signals. This is called Common Channel Interoffice Signaling (CCIS). CCIS is illustrated in *Figure 1-13*. The basic control is by digital computer and CCIS is a separate data network for exchanging control signals among these computers. As the name suggests, this method of signaling is used on the interconnecting trunks that carry signals between central offices.

Figure 1-13.
Common Channel
Interoffice Signaling
(CCIS)

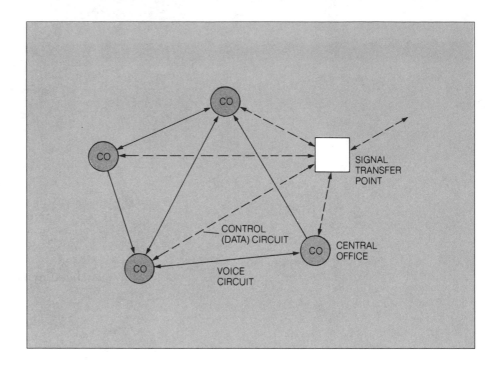

Digital Transmission

Thanks to the sophistication of solid-state technology, voice/analog signals may be digitized before being sent over the telephone lines. This permits time division multiplexing (TDM) and pulse code multiplexing (PCM), allowing many control and voice signals to be combined.

Many advances have been made in solid-state electronics and integrated circuits that handle digital signals, so that high functional density integrated circuits with expanded signal handling capability can be put into a small space, are low cost, operate with low power, and have long term reliability. Because of this, telephone system designs are changing toward an all digital network. The newer systems convert voice and signaling information to digital signals for transmission. When both voice and signaling information are transmitted in digital form it is called digital transmission.

Instead of the voice signal being processed as an analog signal, it is converted into a digital signal and handled with digital circuits throughout the transmission process. When it arrives at the central office that serves the called telephone, it is converted back to an analog signal to reproduce the original voice transmission. (In the future, the digital signal may travel all the way to the telephone set.)

When the binary signal is transmitted in serial form as shown in *Figure 1-12,* and the code varies as the signal changes, the method is called pulse code modulation (PCM). Because the voice signals and the supervisory signals are low-frequency signals (300–3,000 Hz), and digital circuits can operate at very high frequencies (millions of cycles per second), voice signals from many conversations can be sent in series on the same line. This is called time division multiplexing (TDM).

For digital transmission, multiplexing is done in a particular way as shown in *Figure 1-14*. To illustrate the technique, suppose a person is located at point A and can see the binary codes of *Figure 1-14*. The binary codes pass by point A serially one bit at a time. In this case, there are eight bits (d_0 to d_7) in each code. The value of the signal on Channel 1 is represented by the combination of 1s and 0s in the 8-bit code for Channel 1. The value of the signal on Channel 2 is represented by the combination of 1s and 0s in the 8-bit code for Channel 2. Channel 2's code follows Channel 1's code. Instead of the transmitted signal continuously representing the signal of one channel as on an analog circuit, the signals from all 24 channels are mixed together, but in a definite pattern. Channel 2 is multiplexed behind Channel 1, Channel 3 behind Channel 2, and so on until the codes for 24 channels have been multiplexed together, one following the other in serial fashion in time as shown. So the observer at point A would see 8-bits for Channel 1, 8-bits for Channel 2, 8-bits for Channel 3, and so on until Channel 24 came by. The pattern would then repeat beginning again with Channel 1. This is the way that many conversations that have been digitally encoded by PCM are digitally transmitted by TDM over one channel. TDM and PCM will be covered in much more detail in later chapters.

Figure 1-14.
Multiplexed PCM
Signals *(Source: D.L. Cannon and G. Luecke,* Understanding Communications Systems, *SAMS, A Division of Macmillan Computer Publishing, 1984)*

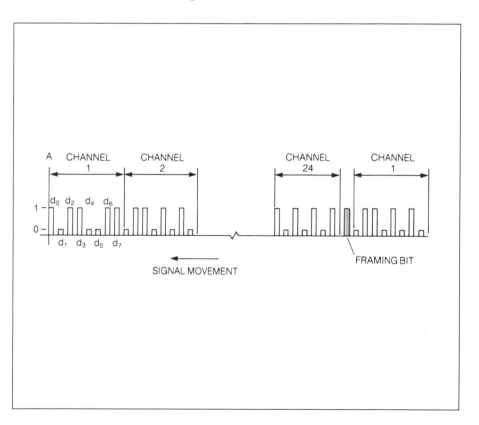

SWITCHING SYSTEMS

Now that there is some understanding of the transmission of the signals, let's look at how the called telephone and the calling telephone are actually connected by selecting one of a multitude of paths. Several mechanisms have been used to provide control of the process of switching one circuit among many others in the hundred-odd years since Mr. Bell envisioned "this grand system." The first, of course, was manual.

Manual Control

Early switchboards connected two telephones together by manually patching them together at the switchboard console.

Early telephone switchboards were operated manually using a jack for each line and two plugs on a long flexible wire, called a cord pair (*Figure 1-15a*), for making the connection. The cord pairs appeared in rows on a shelf in front of the operator, and the jacks (called line appearances) were mounted on a vertical panel as shown in *Figure 1-15b*. To make a connection, the operator picked up a cord (*Figure 1-15c*), plugged it into the jack corresponding to the line requesting service, obtained from the calling party the name or number of the desired party, then plugged the other end of the cord pair into the correct outgoing line jack. There are many thousands of cord switchboards still in operation, a tribute to the versatility and ease of programming of the control system, and the personal touch it provides. However, not every subscriber appreciated the personal touch.

Progressive Control

In 1889, a Kansas City, Missouri undertaker Almon B. Strowger, began to suspect that potential clients who called the operator of the local manual exchange and requested "an undertaker" were more often than not being connected to a firm down the street. This suspicion was reinforced when he learned that the telephone operator was the wife of the owner of the other funeral parlor in town. In the best tradition of pioneer America, Mr. Strowger invented a mechanical substitute for the biased operator which could complete a connection under direct control of the calling party. This simple device is variously called the Strowger, two-motion, or step-by-step switch. It was patented in 1891 and became the basis of a very large fraction of the installed telephone switching systems in the world. As of 1978, 53% of the Bell System exchanges in service (over 23,000,000 subscribers) used Strowger switching, even though the Bell System did not begin installing Strowger switching until about 1918.

Step-by-Step

The Strowger, or step-by-step, switch connects pairs of telephone wires by progressive step-by-step operation of several series switches (called the switch train) operating in tandem. Each operation is under direct control of the dial pulses produced by the calling telephone. *Figure 1-16* shows what happens in simplified schematic form. The telephone

Figure 1-15.
Manual Switchboards

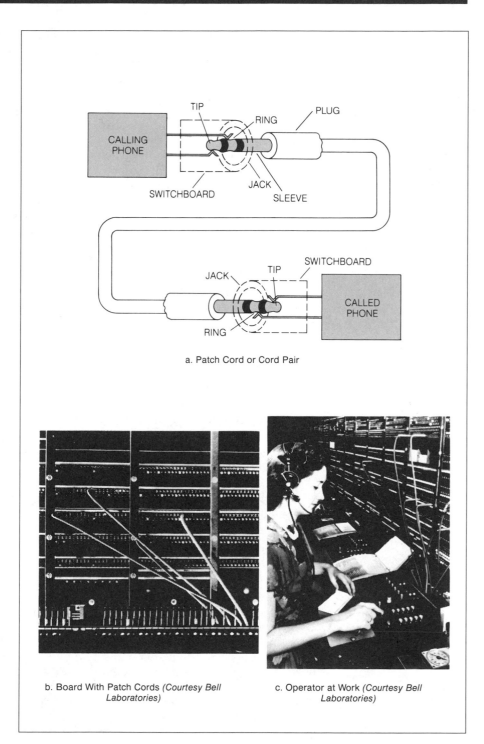

a. Patch Cord or Cord Pair

b. Board With Patch Cords *(Courtesy Bell Laboratories)*

c. Operator at Work *(Courtesy Bell Laboratories)*

The Strowger device responds to the dial pulses generated by the calling telephone. The two-wire connection to the receiving telephone is made by a series of sequential operations of the switch train operating in tandem. The switches step vertically for each dial pulse, and then horizontally to find a free selector position.

line shown is actually a pair of wires. When the calling telephone goes off hook, current flowing in the local loop operates a relay in the exchange, causing the first switch in the train (the linefinder) to search for the active line by stepping vertically until the vertical contact is connected to the off-hook line. The linefinder then steps horizontally until it finds a first selector that is not in use on another call. This is the next switch in the train. When a free first selector is connected, a dial tone is returned to the calling party. The first selector switch waits for the first digit to be dialed, then steps vertically one step for each dial pulse received. When it has taken in one digit, it steps horizontally until a free second selector is found, and the process is repeated. Thus, the first switch in the train (the linefinder) takes in no digits, the second and third takes in one digit each, and the last switch in the train (called a connector) takes in the final two digits. A 10,000 line exchange requires four digits to be dialed (0000 through 9999) and requires four switches for each connected call (linefinder, first selector, second selector, and connector). A line drawing of a Strowger switch bank is shown in *Figure 1-17*.

The Strowger switching system has the following significant limitations:

1. Since several switches are operated in tandem and the switches (except for the first one) are shared among many incoming lines, it is possible for a call to become blocked part way through the dialing sequence, even though the called line is free.

2. It is not possible to use tone dialing (DTMF) telephones directly. (They may be used if the central office is equipped with a conversion device.)

3. The switch requires the successful sequential (step-by-step, time related) operation of several relays and a sizable voltage and current is switched each time a switch is stepped. Consequently, the mechanical reliability of the switches is low, they require large amounts of maintenance by skilled people, and they generate large amounts of electrical and mechanical noise.

4. Since the switching network is hard-wired, it is difficult to make changes in the switching arrangement.

Common Control

Common control provides more sophistication and flexibility in the way calls are routed.

Primarily due to the inflexibility and large maintenance costs associated with Strowger type switching networks, the concept of common control was brought back, but with a new type of switching matrix called a crossbar. The common control can be assigned to an incoming call as required. It takes in the dialed digits, and then sets up the path through the switching matrix according to hard-wired or stored-program rules. These rules provide for variations in the handling of local and long distance calls, for choosing an alternate route for a call in case the first route chosen is busy, and for trying the call again automatically in case of blocking or faults in the switching path. The common control element may be a relay operated device called a marker or a stored-program controlled digital computer.

**Figure 1-16.
Strowger Step-by-Step
Switching**

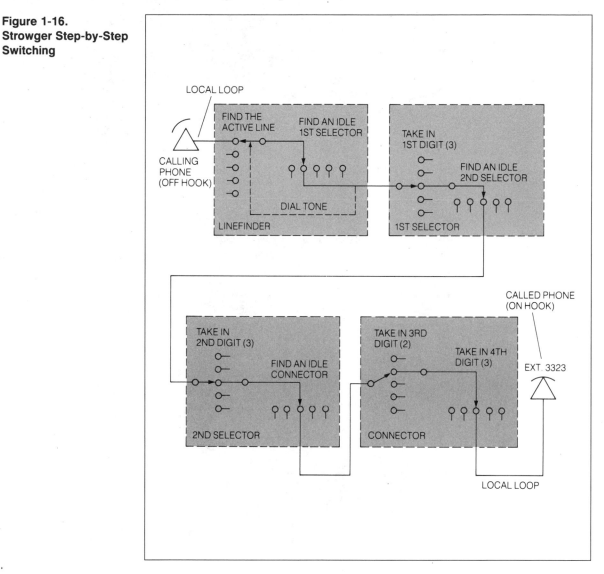

Figure 1-17.
Strowger Switch Bank
*(Courtesy Bell
Laboratories)*

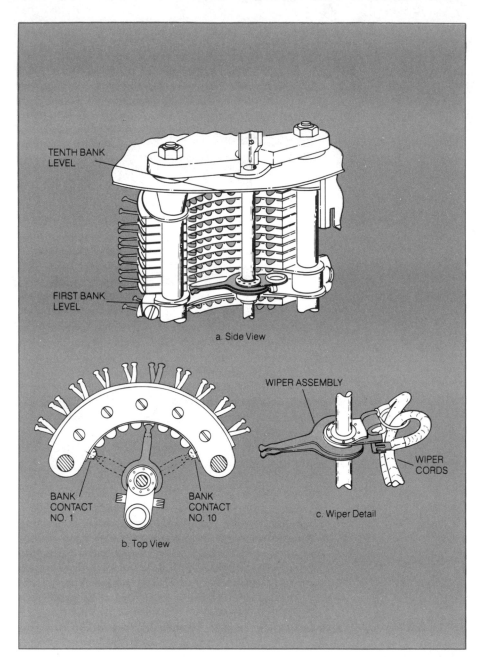

TENTH BANK
LEVEL

FIRST BANK
LEVEL

a. Side View

WIPER ASSEMBLY

WIPER
CORDS

c. Wiper Detail

BANK
CONTACT
NO. 1

BANK
CONTACT
NO. 10

b. Top View

Crossbar

The crossbar system utilizes a switching matrix, which is externally managed by common control, to route telephone calls. Its operation depends on a connection made by energizing a vertical line and a horizontal line in the matrix.

Crossbar, as the name implies, depends on the crossing or intersection of two points to make a connection. The switching matrix is shown in *Figure 1-18a*. It is called a crosspoint array. Its operation depends on energizing a vertical line and a horizontal line and the point where they intersect represents the connection made. Therefore, any one of the input lines shown (I1 through I6) can be connected to any one of the output lines (O1 through O10) by energizing a particular input line and a particular output line.

As shown in *Figure 1-18b,* the crossbar matrix is controlled by common control. Control signals from transmission lines are detected and used to control the matrix to connect the proper lines for the path from the calling telephone to the called telephone.

Electromechanical Version

An electromechanical version utilizes electromagnets to open and close contacts in the matrix. After many operations, the contacts may prove to be unreliable.

Figure 1-19 illustrates a relay type of mechanism that has vertical and horizontal selector bars operated by electromagnets which close relay contacts to provide the matrix interconnection. When a horizontal select magnet is energized, its horizontal selecting bar rotates slightly on its axis. This moves a selecting finger up or down to allow either an upper or lower bank of horizontal contacts to complete a circuit to the vertical contacts when the appropriate vertical select magnet is energized. The vertical select magnet moves the vertical holding bar sideways to push on the selecting finger to close the respective horizontal contacts to the vertical contacts. For the other horizontal select magnets that have not been energized, the associated selecting fingers are in the middle position and pass between the horizontal contacts when the vertical holding bar pushes on the selecting fingers. Therefore, no crossbar connection is made at any other point.

One crosspoint is at the intersection of each horizontal and vertical bar, as shown in *Figure 1-18a*. A crosspoint is provided for each wire of the wire pair so that both wires of the line are switched. Once connected, the switch path is maintained by the current flowing through the vertical select magnet coil. The horizontal select magnet is deenergized and the horizontal selecting bar returns to its idle position, but the previously selected finger is held by the vertical holding bar. When the calling end goes on hook (hangs up), the vertical select magnet current is interrupted and the crosspoints are released. Thus, the calling party controls the connection.

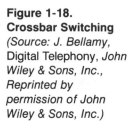

Figure 1-18.
Crossbar Switching
(Source: J. Bellamy, Digital Telephony, *John Wiley & Sons, Inc., Reprinted by permission of John Wiley & Sons, Inc.)*

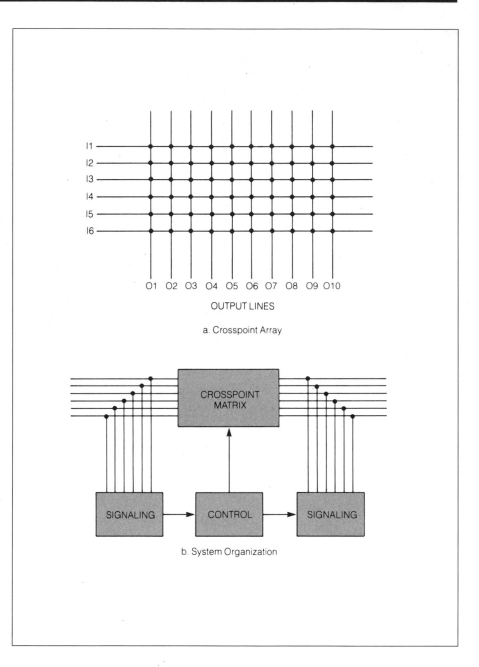

a. Crosspoint Array

b. System Organization

Figure 1-19.
Crossbar Switch
(Courtesy TeleTraining,
Inc. Geneva, IL)

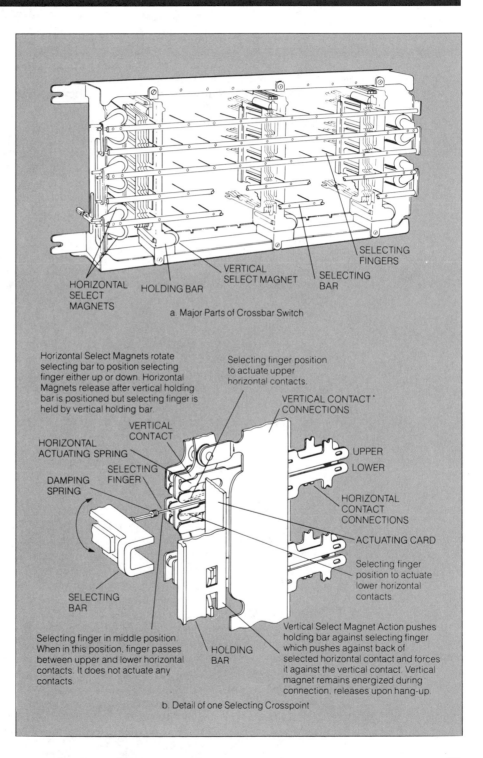

HORIZONTAL
SELECT
MAGNETS

HOLDING BAR

VERTICAL
SELECT MAGNET

SELECTING
BAR

SELECTING
FINGERS

a. Major Parts of Crossbar Switch

Horizontal Select Magnets rotate
selecting bar to position selecting
finger either up or down. Horizontal
Magnets release after vertical holding
bar is positioned but selecting finger is
held by vertical holding bar.

Selecting finger position
to actuate upper
horizontal contacts.

VERTICAL CONTACT
CONNECTIONS

HORIZONTAL
ACTUATING SPRING

VERTICAL
CONTACT

SELECTING
FINGER

DAMPING
SPRING

UPPER

LOWER

HORIZONTAL
CONTACT
CONNECTIONS

ACTUATING CARD

Selecting finger
position to actuate
lower horizontal
contacts.

SELECTING
BAR

Selecting finger in middle position.
When in this position, finger passes
between upper and lower horizontal
contacts. It does not actuate any
contacts.

HOLDING
BAR

Vertical Select Magnet Action pushes
holding bar against selecting finger
which pushes against back of
selected horizontal contact and forces
it against the vertical contact. Vertical
magnet remains energized during
connection, releases upon hang-up.

b. Detail of one Selecting Crosspoint

Reed Relays

Reed relay switches, although also electromechanical devices, are more dependable because they are in a sealed envelope. They open or close depending on the polarity of the electrical impulses inputted.

Another type of switch uses reed relays to make the connections. The reed relay is a small, glass-encapsulated, electromechanical switching device as shown in *Figure 1-20*. These devices are actuated by a common control which selects the relays to be closed in response to the number dialed, and sends pulses through coils wound around the relay capsules. The pulses change the polarity of magnetization of plates of magnetic material fitted alongside the glass capsules. The contacts open or close in response to the direction of magnetization of the plates, which is controlled by the positive or negative direction of the pulse sent through the windings. Since the contacts latch, no holding current is required for this type of crosspoint, but separate action is required by the common control to release the connection (unlatch or reset the relay) when one party or the other hangs up.

Reed relays have improved the reliability and maintainability of switches a great deal. Crossbar switches still provide much of the switching for long distance or long haul telephone calls in the United States. In addition, reed relays are an important part of stored program controlled electronic switching systems.

All of the step-by-step, crossbar, and reed relay switching is called space division switching because each telephone conversation is assigned a separate physical path through the telephone system. The PCM time division multiplexed digital transmissions discussed previously are different because they place many interleaved conversations onto one telephone line.

Local Loops and Trunks

As the telephone connection extends from the calling phone, it proceeds over the local loop to the first exchange, the central office. From the central office, it proceeds over trunk connections selected by the control signals to other exchanges. Control signals and functions different from those used in the local loop are required for making trunk interconnections. Let's examine several of these.

When control and function signals are interchanged between local and switching office and the higher level trunks, their formats must be changed. It is vital, however, that the data itself must remain intact so that the normal phone functioning continues.

A dial tone is returned to the calling telephone when the local exchange is ready to receive the called number after recognizing that a telephone has been taken off-hook. To tell a trunk the same thing requires an appropriate trunk acknowledgement signal.

The acknowledgement signal may be a reversal of battery polarity, momentary reapplication of a tone, an interruption of the circuit (wink), or transmission of coded tone signals (proceed-to-send tone). The acknowledgement signal is returned only after the exchanges have made the proper connections to receive the called number.

The called number must be transmitted over the trunk just like the dial pulses or dial tones from the calling telephone. This is done with a sending device in the originating exchange. At the receiving exchange, the common control may be the crossbar control circuit, or (on newer exchanges) an electronic circuit called an incoming register that receives

**Figure 1-20.
Reed Relay Structure**

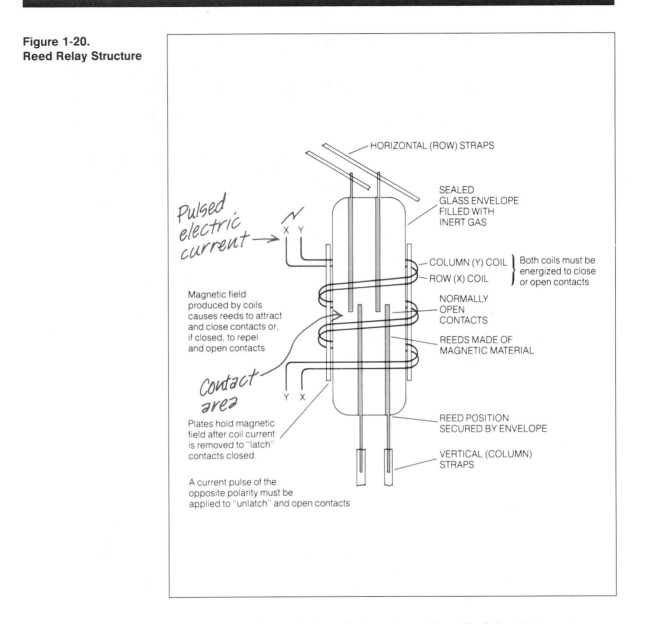

and stores the digits of the called number. After all of the digits arrive, they are used by the common control section to direct the switching of the network to route the conversation over the correct path.

When an exchange has received enough information to complete a connection, it selects the appropriate outbound trunk or line, seizes it, and applies a signal to alert the distant end that a call has arrived and requires service. When the next exchange detects the call, it sends back an acknowledgement signal to the sending exchange to send the dialed number.

At the same time that the call connection is progressing toward the called phone, each exchange in the train sends back status signals to the previously connected exchange to indicate how the call is progressing.

When the call arrives at the local central office serving the called telephone, and if the called telephone is on-hook, the ringing voltage is sent over the local loop to ring the called telephone. At the same time, the central office serving the called telephone returns a ringback signal all the way back to the calling phone to indicate that the called telephone is ringing.

When the called telephone answers, its central office provides the power for the called phone and the conversation proceeds.

TRANSMISSION SYSTEM FACILITIES

There are three categories of transmission facilities: metallic, analog carrier, and digital carrier.

The physical lines, the relays, switches, cables, power supplies, electronic circuits, transformers, impedance matching networks, etc., are called the transmission system facilities. Whether voice only, voice and analog signaling, digital signaling, or all digital signals are sent over a particular path depends on the facilities. Also, the facilities determine the number of channels that can be sent over a particular path. Facilities generally fall into three categories: Metallic, Analog Carrier, and Digital Carrier.

Metallic

A metallic facility is the simple twisted wire hook-up between the telephones and the central office.

Metallic facilities are wire pair line circuits carrying only one VF channel, and on which in-band or out-of-band signaling, including dc signaling, can occur. Typically, a metallic facility is used to connect a central office to the telephones in homes and small businesses.

Analog Carrier

Analog signal carriers may be classified as solid conductors, which include twisted pair open wire, twisted pair in cable, coaxial cable pair, and fiber optic cable or wireless including land microwave and satellite.

Analog carrier facilities may operate over different media, such as wire lines, multiwire cable, coaxial cable, microwave radio land or satellite links, or fiber optic cable. Each circuit carries from a few to several thousand individual analog VF channels and in-band analog or CCIS signaling is used.

Table 1-5 shows the common types of analog carrier facilities, their bandwidth and the number of voice channels that can be carried by each. Note that an L5 transmission path can carry 108,000 different conversations at the same time.

One interesting application of analog multiplexing occurs in undersea telephone cables. The current undersea cable, called the SG system (TAT-6; *Trans-Atlantic Telephone system 6*), carries 4,000 channels in both directions on a single coaxial conductor 1.7 inches in diameter. Repeater amplifiers are placed approximately every 5 nautical miles, and equalizers are placed about every 100 to 150 nautical miles. The electronic repeater amplifiers and equalizers are powered from one end by

**Table 1-5.
Analog Carrier
Facilities**

Medium	Carrier Type VF	Number of Channels	Number of Pairs or Radio Channels	Bandwidth
Open-Wire	O	4–12	2	200 kHz
	On-2	24	2	200 kHz
Twisted Pairs in Cables	K	12	2	300 kHz
	N-1	12	2	300 kHz
	N-2	12	2	300 kHz
	N-3	24	2	300 kHz
	N-4	24	2	300 kHz
Coaxial Cable Pairs	L1	1,800	3	3 MHz
	L3	9,300	5	10 MHz
	L4	32,400	9	20 MHz
	L5	108,000	10	68 MHz
Microwave Radio	TD-2	19,800	11	500 MHz
	TD-3	12,000	10	500 MHz
	TH-1	10,800	6	500 MHz
	TH-3	14,400	6	500 MHz
	TM-1	3,600	4	500 MHz
	TJ	1,800	3	1000 MHz
	TL-1	720	3	1000 MHz
	TL-2	2,700	3	1000 MHz
	AR6-A	42,000	7	500 MHz

a 7,000 volt power supply which also feeds over the center conductor. The system operates on two bands: 1.0 to 13.5 MHz and 16.5 to 29.1 MHz.

Fiber optic cables are currently replacing the older coaxial cables of the SG (TAT-6) system. The optical fiber used in the new TAT-9 system will handle more than 100,000 telephone channels simultaneously.

Digital Carrier

Digital facilities are similar to analog, but the information is in a digital format for transfer.

Digital carrier facilities may operate over the same types of media as analog carrier terminals, but all of the voice and control signals are converted to digital data, multiplexed, and sent as a continuous digital data stream over a single channel. *Table 1-6* shows information for digital carrier multiplex systems. Note that the more voice channels handled, the higher the digital signal data rate in bits per second (bps).

CCIS Facilities

The CCIS system it separates the voice channels and control signals into separate channels.

As previously discussed, the Common Channel Interoffice Signaling (CCIS) system operates over a channel with VF bandwidth that carries all the supervisory control signals for several VF channels carrying voice only. The facilities necessary to accomplish this are shown in *Figure 1-21*.

Table 1-6.
Digital Carrier and
Multiplex Systems

Medium	Digital Signal Designation	Multiplex Designation	Number of VF Channels	Data Rate (Mbps)
T1 Paired Cable 1A-RDS Radio	DS-1	D-Channel Bank	2	1.544
T1C Paired Cable	DS-1C	M1C	48	3.152
T2 Paired Cable	DS-2	M12	96	6.312
3A-RDS Radio FT-3 Fiber Optic	DS-3	M13	672	44.736
T4M Coaxial Cable	DS-4	M34	4032	274.176
WT4 Wave Guide	DS-4	M34	4032	274.176
DR18 Radio	DS-4	M34	4032	274.176
FT4 Fiber Optic Cable	DS-4	M34	4032	274.176

Figure 1-21.
CCIS Signaling
Facilities *(Courtesy AT&T)*

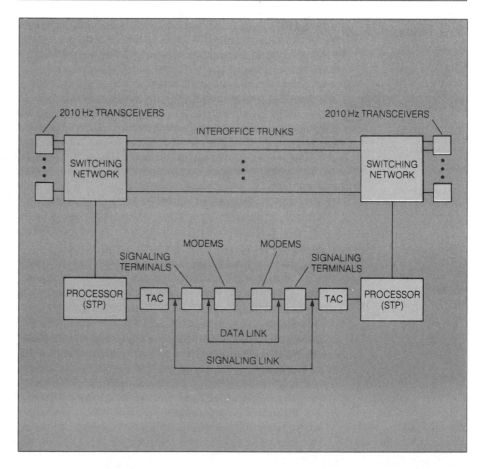

The CCIS link consists of a Terminal Access Circuit (TAC), a signaling terminal, and a modem (discussed in Chapter 6) at each exchange; and a channel to a control computer called a Signal Transfer Point (STP). Several exchanges may be interconnected with two or more STPs for back-up protection against equipment failure.

CCIS is a major improvement in network control. It provides higher signaling speed, much greater information capacity, and two-way signaling even while a conversation is in progress on a trunk. Also, it provides some benefits of separate channel signaling such as freedom from accidental disconnection by voice or extraneous tones that appear to be control tones, protection from misuse of the network by simulating in-band signals, and freedom from mass seizures of whole trunk groups due to carrier or multiplex system failures.

Transmission Mediums

The transmission medium is what actually carries a signal from point to point in the network. The signal carried by the medium may be voice or data, network control signals, or combinations of the two. Each medium has its own particular advantages and disadvantages. Although modern transmission mediums can be found in many shapes and sizes, they can typically be separated into three categories: wire, radio, and fiber optic.

Wire Medium

Twisted pair cable, coaxial cable, and open wire lines are common transmission media found not only in the local loop and exchange network, but in the long haul network as well. Wire is certainly the oldest and most straightforward of all mediums as shown in *Figure 1-22,* yet it still remains the foundation of the network. New technologies promise to replace wire in the next few decades.

Wire has several disadvantages. It is expensive, heavy, and bulky. The cost of installing and repairing long-haul wire is often prohibitive when compared with other new media. Wire is also susceptible to such environmental effects as corrosion, noise, and voltage spikes.

Radio Medium

Radio takes into account microwave and satellite communication. It offers many advantages over wire in the exchange and long haul networks. A typical microwave link can handle more than 40,000 voice channels in an analog system. The network can be altered or rearranged easily without having to relocate huge amounts of copper or fiber optic cable. Maintaining radio systems is also less expensive than a wire network.

**Figure 1-22.
Typical Wire
Transmission Media**

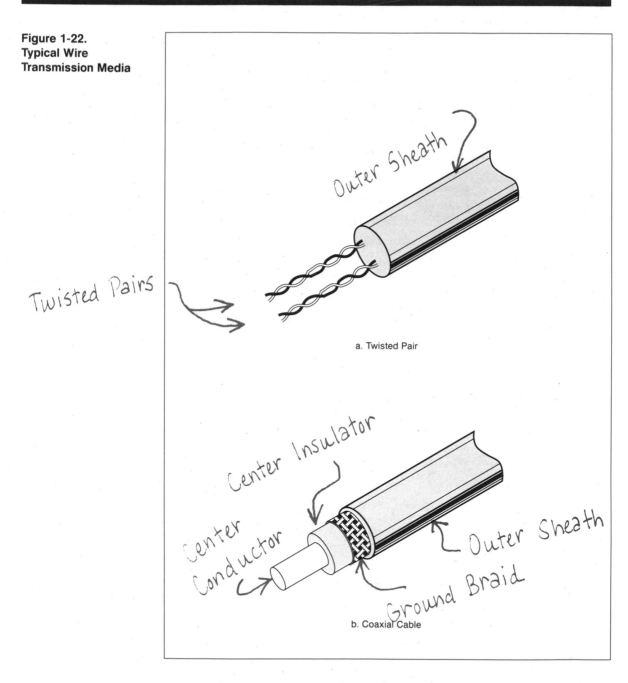

a. Twisted Pair

b. Coaxial Cable

On the negative side, radio system installations require huge capital expenditures for structures, equipment, and real estate. Radio systems are also subject to the problems associated with propagation and atmospheric conditions. Reflection, refraction, diffraction, fading, noise, and interference can all work to degrade the transmission signals through the atmosphere. *Figure 1-23* shows a typical radio relay link.

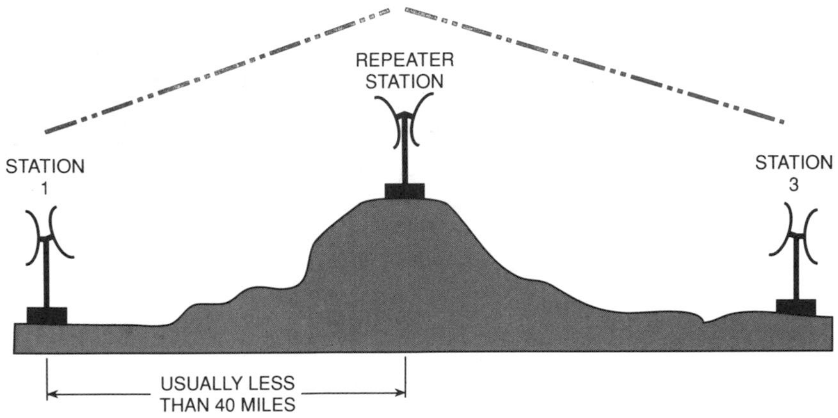

**Figure 1-23.
Typical Radio Relay
Link**

Waveguide bridges the gap between wire and radio. It provides a means to transmit microwave signals through a hollow metal conductor—much like cable. Waveguide offers the advantages of the radio medium, while keeping the signal from ever entering the atmosphere where degradation can occur. It also provides the same bandwidth capacity as open broadcast microwave and contributes little loss to the signal. *Figure 1-24* illustrates two common styles of waveguide.

Unfortunately, waveguide is much more delicate and expensive than wire medium. As a result, its use is limited to very short transmission distances. Any physical damage or corrosion in the waveguide will introduce severe losses to the signal.

Optical Fiber

Optical fibers represent the newest frontier in telecommunication transmission media. These fibers are made by surrounding a high-quality glass core with a glass cladding material of a different refractive index. Plastic core and cladding materials are also commonly used. Light introduced to the core is carried down the fiber by continuously reflecting at the core-cladding interface. *Figure 1-25* presents a diagram of a typical fiber, and shows how light travels through it.

**Figure 1-24.
Waveguide Styles**

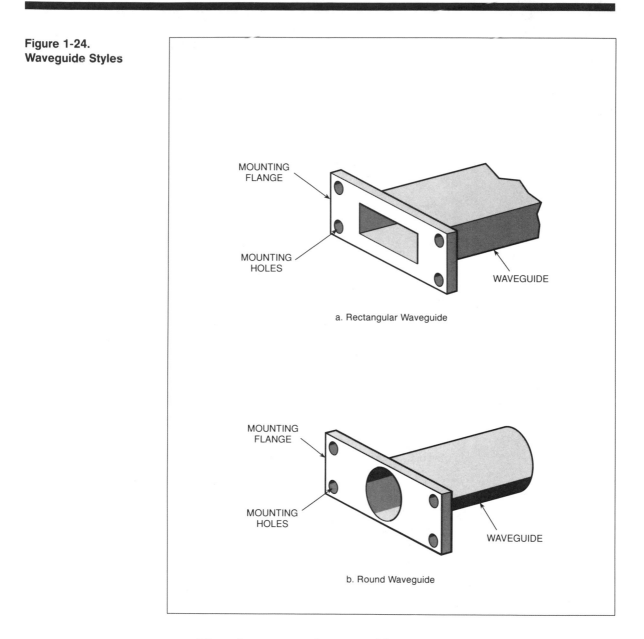

MOUNTING
FLANGE

MOUNTING
HOLES

WAVEGUIDE

a. Rectangular Waveguide

MOUNTING
FLANGE

MOUNTING
HOLES

WAVEGUIDE

b. Round Waveguide

Fibers have many advantages. They are very thin and light, but are surprisingly strong. A typical fiber can carry signals for great distances with less than 1 decibel per kilometer of attenuation. Repeater equipment needs to be placed only every 10 miles or so. Fibers have a wide signal bandwidth—some systems carry over 30,000 voice channels simultaneously. They carry no electrical current and are free from noise, ground loop effects, crosstalk, and interference. This makes optical fiber the ideal choice for long haul installations.

Figure 1-25.
Typical Optical Fiber

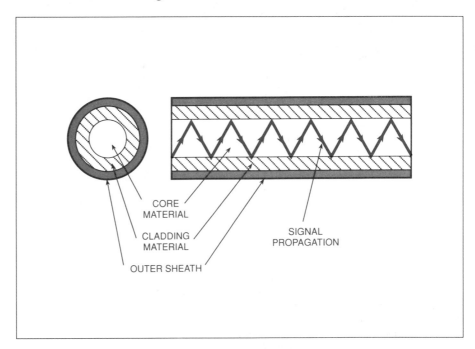

CORE
MATERIAL

CLADDING
MATERIAL

SIGNAL
PROPAGATION

OUTER SHEATH

Yet, optical fibers have several drawbacks at this time. First, building connectors for fibers is difficult for high-reliability, rugged environmental conditions. Second, making multiple taps from the same piece of fiber is next to impossible, so its use is usually limited to direct point-to-point applications. In spite of these difficulties, however, optical fiber continues to grow in capacity and flexibility.

SYSTEM OPERATING CONDITIONS

Telephone systems have certain specifications that are used as guidelines throughout the industry. Here are some of these.

Telephone Set

Telephone set electrical parameters have been standardized by the industry.

The telephone set is designed to operate under a wide range of electrical, mechanical, and acoustical conditions. Some of the design parameters are dictated by human factors such as sound pressure levels, handset dimensions, etc.; some are historical carryovers such as ringing voltage and frequency; and some, such as minimum line current for satisfactory carbon transmitter and relay operation, are dictated by physical properties of the materials used in the telephone set. *Table 1-7* lists some operating parameters and limits for subscriber telephone sets in the U.S. and in some European countries.

**Table 1-7.
Operating Parameters
and Limits**

Parameter	Typical U.S. Values	Operating Limits	Typical European Values
Common Battery Voltage	−48 V dc	−47 to −105 V dc	Same
Operating Current	20 to 80 mA	20 to 120 mA	Same
Subscriber Loop Resistance	0 to 1,300 ohms	0 to 3,600 ohms	
Loop Loss	8 dB	17 dB	Same
Distortion	−50 dB total	N.A.	
Ringing Signal	20 Hz, 90 V rms	16 to 60 Hz, 40 to 130 V rms	16 to 50 Hz, 40 to 130 V rms
Receive Sound Pressure Level	70 to 90 dBspl*	130 dBspl	Varies
Telephone Set Noise		less than 15 dBrnC**	

*dBspl = dB sound pressure level.
**dBrnC = dB value of electrical noise referenced to −90 dBm measured with C
message weighting frequency response.

TELEPHONE EQUIPMENT REGISTRATION

The FCC has set rules
and guidelines for con-
necting up telephone
devices to the system.

Up until 1955, the FCC supported the monopoly of the common
carriers in the area of terminal equipment and devices attached to terminal
equipment. Terminal equipment is the equipment, such as telephone sets,
that is attached to the network. In 1955, the FCC ruled that the carriers
could not prohibit attachment of devices which were "privately beneficial
without being publicly harmful" to equipment leased by the carrier. (This
was the so-called "Hush-a-Phone" case.) This policy was then extended in
the Carterfone decision in 1968 to apply to equipment connecting to the
switched network. In 1975, the FCC determined that any piece of
equipment registered with the FCC could be connected directly to the
telephone network (this did not include coin telephones and party line
telephones).

The rules for registration of equipment to be connected to the
network are set out in the Rules and Regulations of the Federal
Communications Commission, Part 68: Connection of Terminal Equipment
to the Telephone Network, dated July 1977. This document (available
from the U.S. Government Printing Office) describes the types of
equipment which must be registered, the electrical and mechanical
standards to be met, and standard plug and jack arrangements to be used in
connecting terminal equipment to the network. It also specifies the
procedures to be followed in testing equipment for compliance to Part 68
rules, and the method of applying for certification.

Quiz for Chapter 1

1. The telephone was invented by:
  a. Watson.
  b. Bell.
  c. Strowger.
  d. Edison.

2. The central office detects a request for service from a telephone by:
  a. a flow of loop current.
  b. no loop current.
  c. a ringing signal.
  d. dial pulses.

3. Which kinds of signal are transmitted on the local loop?
  a. voice
  b. tones
  c. pulses
  d. all of the above

4. Which office class is the local central office?
  a. 2
  b. 3
  c. 4
  d. 5

5. Which exchange is used to connect between central offices when a direct trunk is not available?
  a. local
  b. tandem
  c. toll
  d. any of the above

6. Which of the following is a type of dc signaling?
  a. loop current
  b. reverse battery
  c. E & M
  d. all of the above

7. The voice frequency channel pass band is:
  a. 0 to 4,000 Hz.
  b. 300 to 3,000 Hz.
  c. 8,140 to 8,188 Hz.
  d. none of the above.

8. What is used to transmit more than one conversation over a path?
  a. hybrid
  b. tandem
  c. multiplexing
  d. all of the above

9. The common channel interoffice signaling method:
  a. uses the same channel for signaling as for the related conversation.
  b. uses a separate channel for signaling only.
  c. carries the signaling for only one related conversation.
  d. is used on local loops.

10. Telephone switching is accomplished by:
  a. manual switchboard.
  b. step-by-step switches.
  c. crossbar switches.
  d. any of the above.

11. The step-by-step switch:
  a. was invented by Strowger.
  b. generates much noise.
  c. cannot operate directly from DTMF tones.
  d. all of the above.

12. Time division multiplexing is used for:
  a. analog transmission.
  b. digital transmission.
  c. both of the above.

13. Conventional transmission
media include:
   a. twisted pair cable.
   b. microwave waveguide.
   c. fiber optic cable.
   d. all of the above.

The Conventional Telephone Set

ABOUT THIS CHAPTER

In Chapter 1, the telephone set functions were briefly covered to show how a call was initiated and connected. In this chapter, the conventional telephone set functions will be examined in detail. Then, in Chapters 3 and 4, we'll discuss the electronic circuits that perform these functions in electronic telephone sets.

SWITCHHOOK

On-Hook

Figure 2-1 is a block diagram of a telephone set which shows the major functions. The ringer circuit, which will be discussed later, is always connected across the line so it can signal an incoming call. The remainder of the telephone set is isolated from the line by the open contacts of the switchhook when the handset is on-hook. No dc flows (except possibly a small leakage current) since the ringer has a capacitor that blocks dc flow through it.

Off-Hook

In the off-hook condition, the closed switchhook contacts begin a sequence of events in the central office. Initially a relay is energized and a line searcher finds the off-hook line. Then a connection is established, sending a dial tone.

When the handset is lifted off-hook to make a call, the switchhook contacts, S1, S2 of *Figure 2-2a*, close. Loop current flows from the central office battery through the telephone set and through a relay coil at the central office. When sufficient current flows in the relay coil, this relay is energized and its closed contacts signal to other central office equipment that a subscriber telephone is off-hook. A line finder searches until it finds the line with the off-hook signal. The line finder then sets up a connection for the switching equipment to begin receiving the telephone number. At this point, a dial-tone generator is connected to the line to signal the caller to proceed with dialing. Dialing may be done by pulsing (interrupting) the loop current or by sending audio tones. When the first dialed digit is received at the central office, the dial tone is removed from the line.

PULSE DIALING

The rotating dial produces a series of pulsed interruptions in the current flow.

In the conventional telephone set, pulse dialing is accomplished with a rotary dial having ten equally spaced fingerholes as shown in *Figure 2-2a*. The number of dial pulses resulting from one operation of the dial is determined by how far the dial is rotated before releasing it.

The regularly spaced holes in the dial plate and the finger stop make it easy to rotate the dial the correct amount for each digit. This action winds up a spring which rotates the dial back to the rest position when it is released. A small governor inside the dial causes it to return at a constant rate of rotation. A cam turned by the shaft through a rear operates the switch contact S3, shown in *Figure 2-2a*, which opens and closes the local loop circuit during the return rotation of the dial. (The loop is not broken during forward rotation of the dial.) Opening the loop circuit interrupts the loop current flow of 20 to 120 milliamperes and closing the circuit permits the loop current to flow again. Thus, pulse dialing produces a series of current pulses in the loop circuit. One pulse is sent for the digit 1, two or the digit 2, etc., up to ten pulses for the digit 0.

**Figure 2-1.
Telephone Set Block
Diagram**

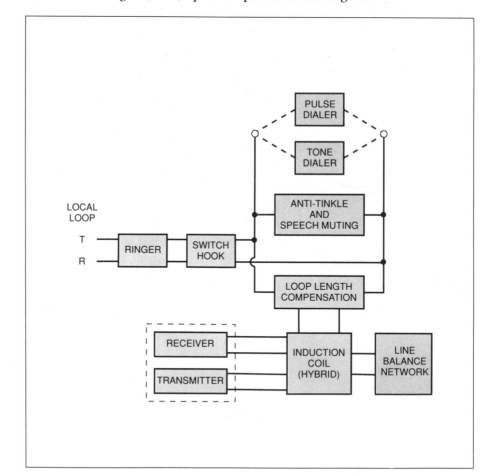

**Figure 2-2.
Dial Pulses**

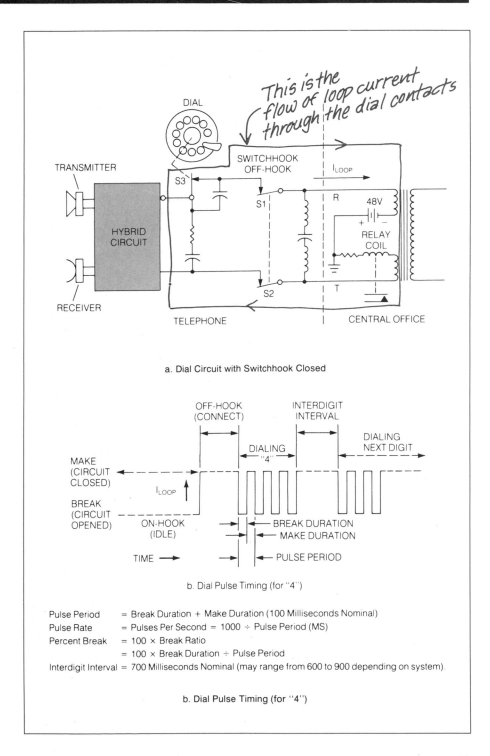

a. Dial Circuit with Switchhook Closed

b. Dial Pulse Timing (for "4")

Pulse Period   = Break Duration + Make Duration (100 Milliseconds Nominal)
Pulse Rate   = Pulses Per Second = 1000 ÷ Pulse Period (MS)
Percent Break   = 100 × Break Ratio
                = 100 × Break Duration ÷ Pulse Period
Interdigit Interval = 700 Milliseconds Nominal (may range from 600 to 900 depending on system).

b. Dial Pulse Timing (for "4")

Pulse Timing

The pulse rate produced by the dial is limited by the relatively cumbersome operation of the dial mechanism and switching relays. The ratio of make (closed) to break (open), time of the dial contacts is 60 percent.

Dial pulses were originally conceived to operate electromechanical switching systems. The mechanical inertia associated with such systems set an upper limit on the operating rate of about 10 operations per second. Thus, mechanical rotary telephone dials were designed to produce a nominal rate of ten pulses per second.

Figure 2-2b shows the timing relations of dial pulses. Note that the number of breaks represent the number dialed. One timed interval of circuit opening and closing, called a dial pulse period, is normally 100-milliseconds long, giving the desired pulse rate of 10 pulses per second. (One second equals 1,000 milliseconds, thus 1,000/100 = 10 pulses per second.) One dial pulse consists of a period when the circuit is open (called the break interval), and a period when the circuit is closed (called the make interval). The nominal value of these periods in the U.S. telephone system is 60 milliseconds break and 40 milliseconds make. This is called a 60-percent break ratio. In other countries, this ratio is usually around 67 percent.

Dial Pulse Detection

The wire pair connecting the telephone set to the central office because of its electrical properties will distort the pulse shape.

For each mile, the wire pair between the central office and the telephone set has shunt capacitance of about 0.07 microfarad, series inductance of about 1.0 millihenry, and series resistance of about 42 ohms which distort the dial pulses in duration and in amplitude. Thus, the dial-pulse detection circuits at the central office must be able to detect dial pulses that are not perfect rectangular-shaped pulses. The circuits also must decode the difference between successive pulses of a digit and the start of a new digit. This is done on the basis of time. As shown in *Figure 2-2b*, the nominal interval between dialed digits (the interdigit interval) is 700 milliseconds.

ANTI-TINKLE AND SPEECH MUTING

When the dial is rotated, the bell and speech circuits are automatically shunted to prevent the dial pulses from actuating them.

High-voltage spikes are produced each time the dial pulsing contacts interrupt the flow of loop current. These spikes of increased voltage can cause the bell of the ringer to sound as the pulses are generated. The ringing is fairly soft, like a tinkle, thus the circuit to prevent it is called an anti-tinkle circuit. This circuit is often combined with the speech-muting circuit as shown in *Figure 2-3*. The extra switches, S5 and S6, are part of a normal dial. When the dial is rotated, the contacts of switches S5 and S6 on the dial close. This shorts the speech circuit to prevent loud clicks in the receiver and possible damage to the speech circuit. The closed contacts also shunt the ringer coil with a 340-ohm resistor to prevent bell tinkle. In this circuit, the ringer capacitor C also serves as a spark quencher to suppress arcing at the dial pulsing contacts.

**Figure 2-3.
Anti-Tinkle and
Speech Muting**

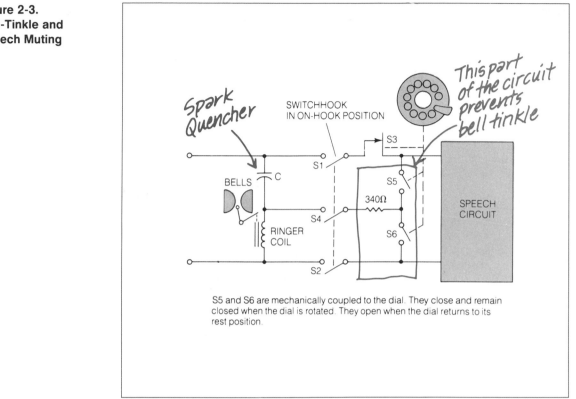

S5 and S6 are mechanically coupled to the dial. They close and remain closed when the dial is rotated. They open when the dial returns to its rest position.

TONE DIALING

In tone dialing, the keypad system utilizes specific pairs of frequencies within the voice band for each key. The method is called dual tone multifreqency (DTMF).

Most telephone sets use the method called dual tone multifrequency (DTMF) for sending a telephone number. These can be used only if the central office is equipped to process the tones. As shown in *Figure 2-4*, instead of a rotary dial, these telephone sets are equipped with a push-button keypad with 12 keys which represent the numbers 0 through 9 and the symbols * and #. (As shown dotted in *Figure 2-4*, some special-purpose telephones have a fourth column of keys for a total of 16 keys.) Pressing one of the keys causes an electronic circuit to generate two tones in the voice band. There is a low frequency tone for each row and a high frequency tone for each column. Pressing key 5, for example, generates a 770-Hz tone and a 1,336-Hz tone. By using the dual tone method, 12 unique combinations are produced from only seven tones when the 12-position keypad is used.

The frequencies and the keypad layout have been internationally standardized, but the tolerances on individual frequencies may vary in different countries. The North American standard is ±1.5% for the generator and ±2% for the digit receiver.

Figure 2-4.
Push-Button Keypad

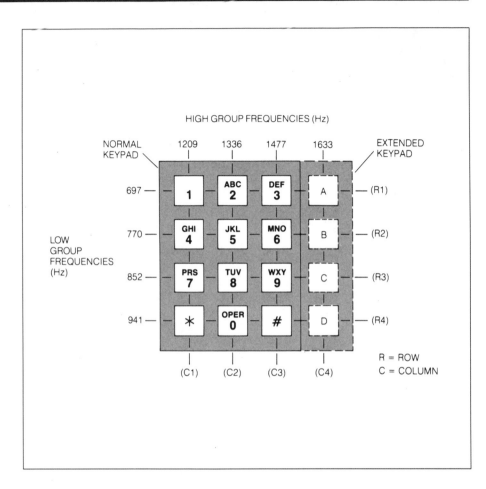

Tone Generation

A dual oscillator circuit generates the two DTMF frequencies simultaneously.

A DTMF circuit using discrete components is shown in *Figure 2-5*. Switches S1, S2, and S3 are shown in the inactive position. With the switchhook in the off-hook position, loop current flows through RV1, L1A, L2A, and via the hybrid network to the line. Transistor Q1 is off. Capacitors C1 and C2 are disconnected at one end by the open contacts of S1 and S2.

When a key is pressed, mechanical couplings, called the row rod and column rod, for that key position close the appropriate S1 and S2 contacts to connect C1 to a tap on L1A and C2 to a tap on L2A. This establishes the resonant circuits for the required low-group tone (L1A–C1) and high-group tone (L2A–C2).

Figure 2-5.
DTMF Circuit

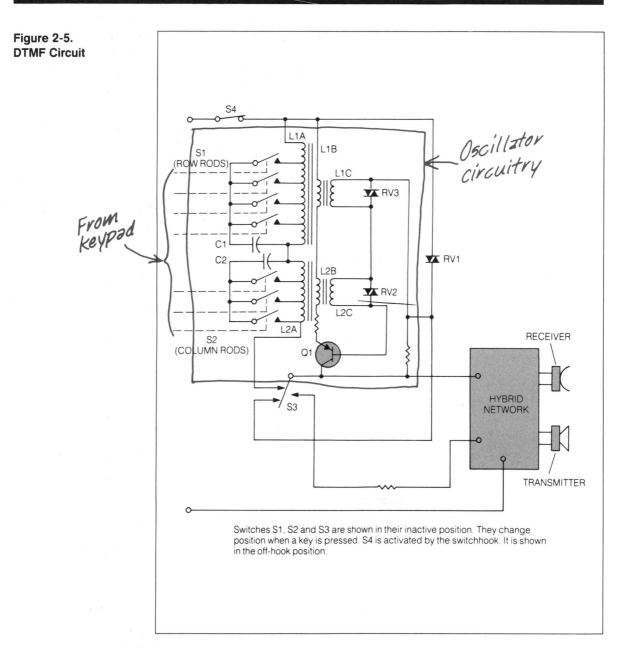

Switches S1, S2 and S3 are shown in their inactive position. They change position when a key is pressed. S4 is activated by the switchhook. It is shown in the off-hook position.

The mechanical arrangement of the push button and switches is such that the resonant circuit connections just described are made with only partial depression of the button. At this point, S3 is still in the position shown. Further depression of the button causes S3 to change position. This interrupts the dc flow through L1A and L2A and shock-excites the two resonant circuits into oscillation. At the same time, S3 connects battery voltage from the line to the collector of Q1. The transformer coupling between L1A, L1B, and L1C, and between L2A, L2B, and L2C cause Q1 to sustain the oscillations and modulate the loop current to transmit the two tones to the central office. The transmitter and receiver are shunted when S3 switches, but the outgoing tones can still be heard in the receiver at a low level. Modern DTMF tones are produced with integrated circuits instead of discrete oscillator components.

Tone Detection

Special frequency and time-selective circuits ensure that only DTMF tones are processed.

The tones used have been carefully selected so that the processing circuits (called the digit receiver) in the central office will not confuse them with other tones that may occur on the line. The digit receiver has frequency selective filters that pass only the frequencies used for DTMF. It also has timing circuits which make sure a tone is present for a specified minimum time (about 50 milliseconds in North America) before it is accepted as a valid DTMF tone.

After a connection to an answered phone has been made, the digit receiver is out of the circuit and the DTMF tones can be transmitted the same as speech. This permits the use of DTMF tones as data communications for entering orders at a remote terminal or obtaining information from a remote data base.

Time Comparison

DTMF is much faster than dial pulsing. At the processing end, this time saving means fewer digit storage registers are required.

DTMF dialing is much faster in principle and in practice than dial pulsing. Using DTMF, the time required to recognize any digit tone is only 50 milliseconds with an interdigit interval of another 50 milliseconds. Thus, the total time to send *any digit* is about 100 milliseconds.

In contrast, dial pulsing requires a 60-milliseconds break and a 40-milliseconds make period for each dial pulse for a total time of 100 milliseconds per *dial pulse*. Thus, for dial pulsing, each higher digit number requires more time because the number of pulses per digit increases. Also, the interdigit interval of about 700 milliseconds (ms) for dial pulsing is much longer. By using the number 555-555-5555, an average time for pulse dialing a long distance call can be obtained as follows:

5 pulses per digit $\times$ 100 ms per pulse $\times$ 10 digits = 5 seconds

Interdigit interval $\times$ (number of digits $-$ 1) = 700 ms $\times$ 9 = 6.3 seconds

Total time for dial pulsing = 5 + 6.3 = 11.3 seconds

DTMF dialing for the same number takes:

Number of digits $\times$ 100 ms per digit = 10 $\times$ 100 ms = 1 second.

These times are the minimum. Actual physical operation of the dial and keypad add to these times, but the time savings from the use of DTMF are substantial when summed for the many telephone calls made in a day. At the local exchange, the digit receiver and memory required to hold the digits as they are keyed are shared among many incoming lines. A reduction in the average time per call that each is in use (called the holding time) means fewer digit receivers are required for the same service. The end result is that less capital equipment is required for a given number of lines; therefore, the cost of the exchange is reduced.

Coupling the DTMF Generator to the Line

There are several important considerations for matching the DTMF generator to the line. It must be able to function within a wide range of variables, including input voltages and line lengths. It must be able to produce acceptable amplitudes, distortion-free signals, and present correct dynamic impedances.

The requirements for the proper interfacing of the DTMF generator to the line are:

1. The correct dc voltages and loop currents must be maintained on any loop length.
2. The tones must have proper amplitude and distortion characteristics.
3. The DTMF generator must have the proper impedance to match the line.

Power
___

Problems may arise when powering the DTMF circuits from the line in either of the two extreme cases—long loops or short loops.

Long loops reduce the amount of current and voltage available for electronic circuits in the telephone; thus, the tone-dialing circuits need to operate from a supply voltage as low as 3 volts.

The minimum working dc voltage ($V_{DC(MIN)}$) for the DTMF generator and interface circuit will be the sum of the peak voltages of the two tones ($V_{LPK} + V_{HPK}$), plus the desired regulated voltage (V_{Reg}), plus the voltage drop necessary to achieve the regulating voltage ($V_{BE} + V_{CE(SAT)}$); e.g.:

$$V_{DC(MIN)} = (V_{LPK} + V_{HPK}) + V_{REG} + (V_{BE}) + V_{CE(SAT)})$$
$$= 1.24\text{ V} + 3\text{ V} + 1.2\text{ V}$$
$$= 5.44\text{ V}$$

Short loops require the telephone set to sink (absorb) high loop current or handle high dc voltage if no provision is made at the central office to regulate or limit the current and voltage supplied to the loop. In either case, the interface must maintain proper dc voltage regulation if it is to supply voltage to other circuits.

Level
___

The sending levels of the DTMF tones are referenced to 0 dBm (1 milliwatt of power dissipated in a 600-ohm impedance). The high frequencies have a level 2 dB above the level of the low frequencies to compensate for losses in transmission. *Figure 2-6* is the curve usually shown to specify the sending levels of the tone pairs.

**Figure 2-6.
DTMF Sending Level
Specification**

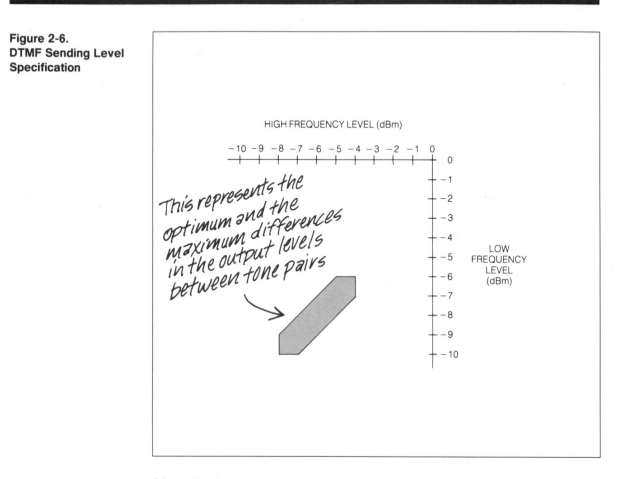

Distortion

Allowable distortion of the transmitted tones is specified in several ways:

1. The total power of all unwanted frequencies shall be at least 20 dB below the level of the frequency pair which has the lowest level.
2. The level of unwanted frequencies produced by the pair shall be:
   a. Not more than −33 dBm in the band from 300 to 3,400 Hz.
   b. Not more than −33 dBm at 3,400 Hz and falling 12 dB per octave to 50 kHz.
   c. Not above −80 dB above 50 kHz.
3. Distortion in dB is defined as:

$$Distortion = 20 \log_{10} \frac{\sqrt{V_1^2 + V_2^2 + --- V_N^2}}{\sqrt{V_L^2 + V_H^2}}$$

where,
    V_1 through V_N are the unwanted frequency components,
    V_L is the low frequency tone,
    V_H is the high frequency tone.

Impedance

When the DTMF generator is active, it must present the correct dynamic impedance to the loop. A nominal 900-ohm impedance is the normal requirement.

When the DTMF generator is inactive, it must present a low dynamic impedance when connected in series with the speech circuit and a high dynamic impedance when connected in parallel.

Return Loss

Return Loss is defined as:

$$RL = 20 \log_{10} \frac{Z_L + Z_g}{Z_L - Z_g}$$

where,
    Z_L is the line impedance,
    Z_g is the output impedance of the telephone set generating the signal.

RL must be greater than 14 dB in the frequency band from 300 to 3,400 Hz and greater than 10 dB in the frequency bands from 50 to 300 Hz and from 3,400 to 20,000 Hz.

Advantages of DTMF

In summary, DTMF dialing is replacing pulse dialing because it:
1. decreases dialing time,
2. uses solid-state electronic circuits,
3. can be used for end-to-end signaling after the call is connected (low-speed data transmission),
4. reduces local exchange equipment requirements,
5. is more compatible with electronically (stored program) controlled exchanges.

TRANSMITTER

The part of the telephone into which a person talks is called the *transmitter*. It converts speech (acoustical energy) into variations in an electric current (electrical energy) that can be transmitted through the transmission system to the receiver of the called telephone. The most common telephone transmitter in use today is in principle like the one invented about 100 years ago by Thomas A. Edison.

Construction

The transmitter converts sound vibrations into equivalent electrical variations. When the vibrating diaphragm compresses the carbon granules, changing their net resistance, a proportional change in the current flow results.

As shown in *Figure 2-7a*, the transmitter consists of a small, two-piece capsule filled with thousands of carbon granules. The front and back are metallic conductors and are insulated from each other. One side of the capsule is held fixed by a support that is part of the handset housing. The other side is attached to a diaphragm which vibrates in response to the air pressure variations caused by speaking into it. The vibrations of the diaphragm vary the pressure on the carbon granules. If the granules are forced together more tightly, the electrical resistance across the capsule decreases. Conversely, if the pressure on the granules is reduced, they move apart and the resistance increases. The current flowing through the transmitter capsule varies because of the varying resistance; thus, the varying air pressure representing speech is converted to a varying electrical signal for transmission to the called party. Other carbon transmitters may be constructed differently, but operate the same way.

Operation

To understand how the level of the loop current varies with the speech level, refer to *Figure 2-7b*. The carbon granule transmitter is represented in the circuit by the variable resistance R_{TR} which changes as a person speaks. The voltage applied across the transmitter is constant; therefore, as the sound energy varies the resistance R_{TR} of the transmitter, the current in the circuit varies in the same manner as the sound intensity. Because the voltage applied is another energy source, the current variations in the circuit deliver electrical energy to the called receiver that produces a greater amount of acoustic energy than that which caused the original resistance variations at the transmitter. Thus, the original speech signal has been amplified.

Effect of Loop Length

The overall resistance of the telephone circuit includes the twisted pair. The longer these lines, the higher the resistance and the lower the circulating current. Since the range of microphone resistance variations remains a constant, the percent of the microphone caused current fluctuations decreases in relation to the steady-state current decrease as the line length increases.

The resistance of the telephone set is typically 400 ohms in the U.S. if measured at points X and Y in *Figure 2-7b* with the local loop disconnected. It is made up of R_{TR} and R_{INT}, where R_{INT} lumps together the rest of the resistance in the telephone besides R_{TR}.

R_L represents the line resistance of both wires. It varies with the length of line to the central office; the longer the line, the larger the resistance. R_B is a balancing resistor in the central office; for this discussion, R_B is considered to be zero.

The loop current is determined by Ohm's law as follows:

$$I_L = \frac{Central\ Office\ Voltage}{R_{TR} + R_{INT} + R_L}$$

The central office voltage in the U.S. is usually 48 volts, therefore:

$$I_L = \frac{48V}{R_{TR} + R_{INT} + R_L}$$

Figure 2-7.
Telephone Transmitter

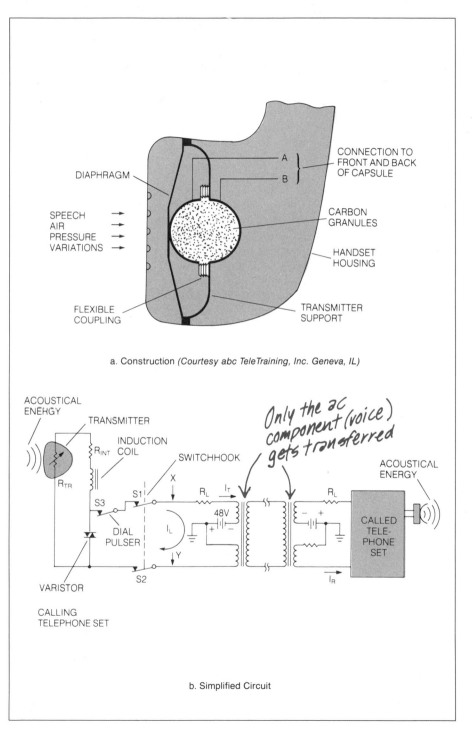

a. Construction *(Courtesy abc TeleTraining, Inc. Geneva, IL)*

b. Simplified Circuit

The total resistance in the local loop ($R_{TR} + R_{INT} + R_L$) gets larger as the line length from the telephone set to the central office increases; therefore, the current gets smaller. Also, as the total resistance gets larger, the variations in R_{TR} as a percentage of the total resistance become much smaller. Therefore, for the same speech input, the variations in the local loop current become smaller as the line length increases. This causes the speech level to go down at the called telephone's receiver. This is not a desirable condition and methods to provide automatic compensation for line length have been developed.

Compensation for Loop Length

A varistor is inserted in the circuit to automatically bypass excess current flow around the transmitter, regulating current flow for a wide range of telephone line lengths.

For ease and uniformity of telephone use, it is desirable that the speech on all calls arrive at the exchange at about the same volume or level regardless of loop length. To achieve this, modern telephone sets have automatic compensation circuits. In *Figure 2-7b*, the varistor resistance decreases as loop current increases with shorter loops. This resistance bypasses some of the current around the transmitter so that the transmitter current is about the same as it is on a long loop. Thus, the varistor automatically adjusts the speech level so that a relatively constant speech level will appear at the exchange regardless of the distance to the telephone set.

Distortion

Carbon transmitters have inherent frequency and dynamic limitations. This results in a distorted replica of the voice.

The principal function of the transmitter is to produce an electrical output waveform that is shaped like the input speech waveform. This output is called an analog signal since the transmitter's electrical output is an analog of (similar to) the acoustic input. Any difference between the waveforms is called distortion. Some forms of distortion are readily apparent to the listener, others are not. The most obvious distortion in the telephone system is the "flattening" quality given to the voice. Some of the distortion is caused by the carbon transmitter and, as shown in Chapter 1, some is due to the fact that some frequencies contained in human speech are not transmitted. This doesn't impair the intelligibility of the speech, but a listener can quickly recognize a recording of a telephone conversation when compared to a recording of someone speaking through a high fidelity sound system. Much of the distortion caused by the carbon transmitter can be eliminated in electronic telephones that have an active speech amplifier by using a low distortion microphone.

Microphones for Use as Transmitters

All microphones may be defined as *electroacoustic transducers*, which convert the varying pressure of a sound wave to a varying current or voltage by means of a mechanical system. Although there are many types of microphones, only two besides the carbon type combine the ruggedness and low cost required for application in telephone sets. One type is the electromagnetic or electrodynamic microphone and the other type is the electret microphone.

Electrodynamic Microphone

The vibrating diaphragm causes a vibrating conductor to cut magnetic flux lines, inducing a proportional voltage in the conductor.

The electrodynamic-type microphone has a diaphragm attached to a wire, coil, or ribbon in a permanent magnetic field as shown in *Figure 2-8*. Sound waves striking the diaphragm cause the coil to move. This movement in the presence of a permanent magnetic field induces a current in the coil proportional to the movement; thus, the sound is converted to a varying electric current which can be amplified and transmitted.

Figure 2-8. Electromagnetic Microphone (Cross Section)

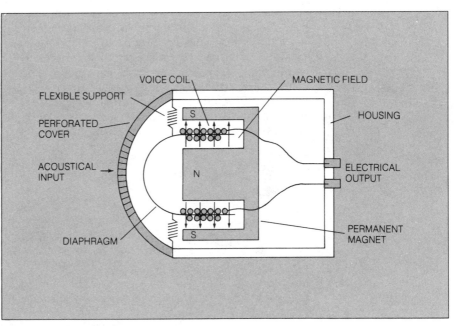

Electret Microphone

The ability of a capacitor to hold a charge is proportional to its capacitance. In the electret microphone, a vibrating diaphragm is one of a capacitor's plates. Therefore, the charge varies in proportion to fluctuations in the sound, appearing as voltage variations at the microphone's output.

The electret-type microphone has been known for many years, although its practical application to telephony awaited development of better dielectric materials and low-cost amplifiers. An electret is the electrostatic equivalent of a permanent magnet. It is a dielectric which will hold an electric charge almost indefinitely.

The relationship of voltage (V), capacitance (C), and charge (Q) is given by the following equation:

$$V = \frac{Q}{C}$$

When an electret is the dielectric between two metal plates, it forms a special kind of capacitor. Charge Q is held permanently in the electret material. Therefore, if one of the plates of the capacitor is the diaphragm of the microphone, the voltage at the terminals will vary according to the diaphragm movement. The variations are very small and must be amplified.

The dielectric material of a modern electret microphone is a fluorocarbon material. It is produced as a foil metallized on one side. The material is charged by placing it in a corona discharge and forcing the electrons into the foil with an external electric field. The resulting film can be assembled into a metal can about 18-mm (0.7 in) in diameter and 18-mm in length.

One type of construction is shown in *Figure 2-9*. The resulting microphone is similar to the studio ribbon-type capacitor microphone, but it does not need a polarizing voltage. Like the ribbon microphone, the device has high internal impedance. Matching to a low-impedance circuit is sometimes done with a source follower FET (field-effect transistor) mounted inside the microphone capsule. The diaphragm has a working diameter of 15 mm (0.59 in) and a capacitance of 15 picofarads. The stability of the charge on the foil is acceptable for commercial use and the total harmonic distortion of the device can be less than 1% compared with 8 to 10% for a carbon granule microphone.

**Figure 2-9.
Electret Microphone
Assembly**

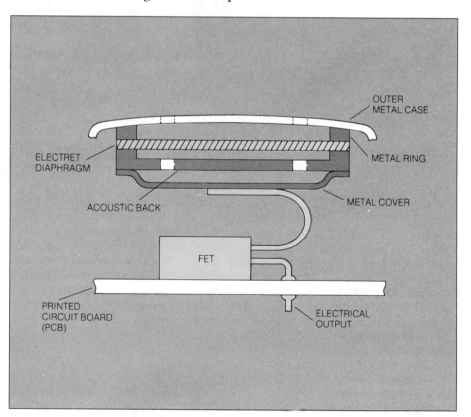

RECEIVER

The receiver transducer produces sound by causing the varying magnetic field induced by the varying sound current to interact with the field of a permanent magnet. The resultant attraction and repulsion vibrates a diaphragm.

The receiver converts the varying electrical current representing the transmitted speech signal (I_T in *Figure 2-7*) to variations in air pressure useable by the human ear. A typical electromagnetic receiver, shown in *Figure 2-10a*, consists of coils of many turns of fine wire wound on permanently magnetized soft iron cores that drive an armature. The armature is a diaphragm made of a soft iron material.

A key requirement for an electromagnetic receiver is a permanent magnet to provide a constant bias field for the varying electromagnetic field to work against; otherwise, both positive and negative currents would push the armature in the same direction. The varying electrical current representing speech (I_R in *Figure 2-7*) flows through the coils and produces a varying electromagnetic field. It alternately aids and opposes the permanent magnet field; thus, it alternately increases and decreases the total magnetic field acting on the diaphragm as shown in *Figure 2-10b*. This causes the diaphragm to vibrate in step with the varying current and moves the air to reproduce the original speech that caused the current changes. (The principle of a permanent magnet providing a bias field also is used in loudspeakers for audio reproduction in many kinds of electronic equipment.)

The receiver shown in *Figure 2-10c* operates similarly except the armature is a separate part and is connected to a conical nonmagnetic diaphragm. The rocking action of the armature causes the aluminum diaphragm to vibrate to reproduce the original speech.

It is interesting to note that the electromagnetic receiver was a central element of Alexander Bell's original telephone patent.

ELECTROMECHANICAL RINGER

When a call has been connected through to the central office serving the called subscriber's local loop, the central office must send a signal to the called party that a call is waiting to be answered. This signaling function is called ringing.

From the telephone company's point of view, it is desirable that a called subscriber answer an incoming call as quickly as possible. While the telephone is ringing, expensive common control equipment in the exchanges, and for toll calls, even more expensive transmission facilities are tied up. When in this condition, these facilities are generating no income. A call arrival alerting system, therefore, must signal incoming calls in an urgent manner and must be loud enough to be heard at considerable distance. For residential phones, it is usually desirable to hear the signal from outside the house. The traditional metallic gong bell struck by a metal clapper fulfills these requirements. It was patented in 1878 by Thomas A. Watson (Mr. Bell's assistant).

**Figure 2-10.
Telephone Receivers**

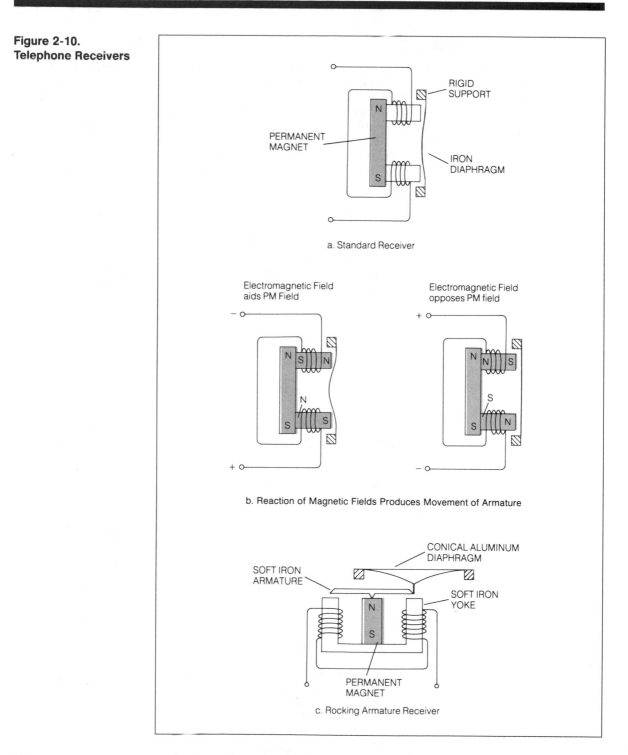

a. Standard Receiver

Electromagnetic Field aids PM Field

Electromagnetic Field opposes PM field

b. Reaction of Magnetic Fields Produces Movement of Armature

c. Rocking Armature Receiver

Operation

The ringer mechanism is actuated by an ac voltage producing a corresponding alternating magnetic field. The field interacts with the permanent magnet fields, causing a pivoted armature to vibrate.

As shown in *Figure 2-11a*, the armature of the ringer is pivoted in the middle and two electromagnets formed by coils wound on permanently magnetized iron cores. These electromagnets alternately attract and repel opposite ends of the armature in response to an applied alternating current. The armature drives a hammer to alternately strike two bells or gongs.

The coils are wound so that the magnetization in the coils due to an applied electric current is of opposite polarity on each of the outer poles of the E-shaped structure. Like the receiver structure, the permanent magnet provides a bias on the armature so that it will be attracted first to one side, then the other, as the applied ac sets up an alternating magnetic field. The permanent magnet also accelerates the hammer as the armature closes so that the hammer strikes the bell harder to produce a louder sound. A mechanical adjustment to regulate the loudness is often included in the telephone set.

A simplified circuit is shown in *Figure 2-11b*. V_R is the ringing voltage applied from the central office. In the U.S., it is usually about 90-V rms at a frequency of 16 to 60 Hz. It produces the alternating current to drive the ringer. Capacitor C passes the ac for ringing, but blocks the flow of any dc from the local loop battery. The value of inductance for the coils and the value of capacitance are chosen so that the ringer circuit presents a high impedance to the voice frequencies. Since the handset is on-hook, the switchhook is open so the ringing voltage is not applied to any other telephone circuits.

Ringing Generator

The voltage source for the alternating current to operate the ringer was at first supplied by a hand-cranked magneto, also patented by Mr. Watson in 1878. The magneto produced about 75 volts at about 17 Hz. The high voltage was necessary due to the inefficiency of existing magnetic materials and system transmission lines. Although the ringer in modern phones is more efficient because of improvement in materials and manufacturing methods, the high voltage still is used today.

The ac ringing signal is generated at the central switching office. Multiple ringing machines run simultaneously; their pauses are staggered. This helps to more evenly distribute the load.

In most central offices, the ac ringing signal is generated by a dc motor driving an ac generator, or by a solid-state inverter. The inverter produces an ac voltage from a dc voltage without any moving parts. Both are called ringing machines and are powered by the −48-V dc central office supply. Since many lines may need ringing at the same time, several ringing machines are used to distribute the load. Usually, these machines are operated so that the ringing times between machines is staggered. By doing this, at least one of the machines will be in the ringing portion (rather than the silent portion) of the cycle at any time. Thus, any new call will have an active ringing signal available immediately rather than having to wait until a silent period ends.

Figure 2-11.
Telephone Ringer

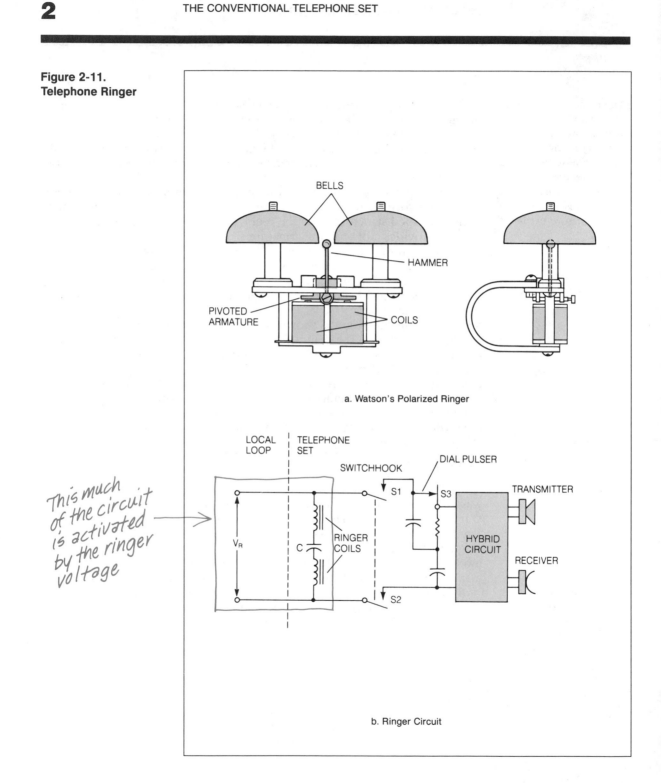

a. Watson's Polarized Ringer

This much of the circuit is activated by the ringer voltage

b. Ringer Circuit

The ac signal is applied to the loop in timed On and Off intervals to produce a ringing cadence. *Figure 2-12* shows two typical cadences. The first is widely used in the U.S. and Europe, and the second is used in the U.K. In some private systems (PABXs), different cadences are used to indicate whether the source of a call is internal or external.

Figure 2-12.
Ringing Cadence

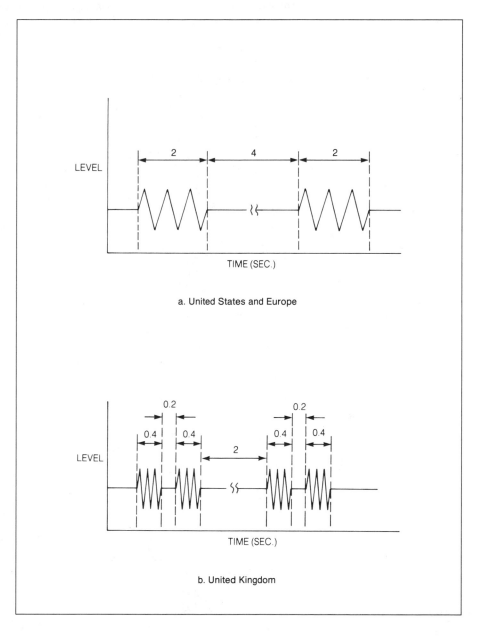

a. United States and Europe

b. United Kingdom

Loop Current Detection in the Presence of Ringing

DC loop current detection is used to determine when the receiver has been lifted, so that the ringing signal may be stopped. The dc current may be applied continuously or between ac ringing signals.

When the telephone is being rung, an alternating current flows in the loop, but no dc flows. If the call is answered while the bell is ringing, it is necessary for the exchange to detect that direct current has begun to flow at the same time that ringing current is flowing, so that the ringing can be removed before the receiver is applied to the called party's ear (this time period is normally considered to be 200 milliseconds). This process is called "ringtrip."

The value of the current in the local loop varies widely. It depends on the exchange battery voltage, the current limiting resistance inserted in the loop at the exchange to limit short circuit current (ranges from 350 to 800 ohms), the resistance of the local loop (which may vary from 0 to about 1900 ohms), and the resistance of the telephone set itself (between 100 and 400 ohms).

The circuits in the central office can detect loop direct currents in the range of 6 to 25 milliamperes as indicating an off-hook condition. However, the detection circuit must not be so sensitive that leakage currents between the tip and ring sides of the loop, or leakage to ground from either conductor, is interpreted as an off-hook condition. The problem is further complicated by the fact that a typical residence may have several telephones all connected to the same line. Since all phones are rung at the same time, the resulting ringing current may reach 50 mA ac, while the loop direct current at the time the call is answered may be lower than 20 mA dc. Since during one half-cycle of the ac wave the central office can't tell dc from ac, ring trip circuits must average the sensed current over at least one full cycle of the ac ringing wave to determine if an off-hook condition exists at the called phone.

The ringing method discussed in the preceding has ringing current applied to the loop superimposed on the dc battery voltage. Another way is to apply the ringing current as pure ac with the dc restored in between bursts of ringing current. The advantage of the first method is that the answer condition can be detected even during the ringing cycle by observing the average dc level. Detecting answer during ringing when only ac is applied is more difficult. This method requires detection of the impedance change when the handset is lifted.

THE HYBRID FUNCTION

Figure 2-13 shows how the transmitter and receiver are connected within the telephone set. It also shows simplified connections at the central office exchange. In particular, note the components labeled "induction coil" and "hybrid." The induction coil name is another of the carryover terms from the early days, but one of its functions in the circuit is similar to that of the hybrid.

**Figure 2-13.
The Hybrid Function in
the Telephone and
Exchange**

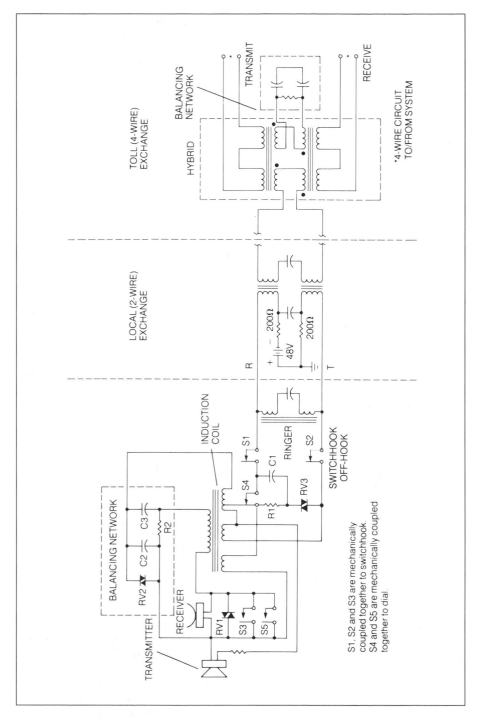

The hybrid circuit permits the interfacing of a two-wire with a four-wire system. Normal full-duplex operations can still maintained.

The function of a hybrid is to interface a two-wire circuit to a four-wire circuit to permit full-duplex operation. In communications, full-duplex means that transmission of signals over the circuit can occur in both directions at the same time. Two-wire circuits are used for the millions of subscriber local loops because they are cheaper. Four-wire circuits, which have two wires for each direction, are used for almost all other circuits in the network. Physical separation of the transmit and receive signals is needed on these circuits so that all types of transmissions can be handled—digital data as well as voice—and so that electronic amplifiers can be used in the transmission path to increase the signal level. A hybrid is used at the central office to interface the local loop to trunks and between trunks at some older offices where only two-wire switching is available.

The telephone handset also is a four-wire circuit with two wires used for the transmitter connection and two for the receiver connection. This is shown in *Figure 2-11* and *2-13*. It can be represented by the circuit as shown in *Figure 2-14*; however, the telephone circuit really evolved from an anti-sidetone circuit rather than a two-wire to four-wire hybrid. For this reason, the circuit in the telephone set more commonly is called an induction coil and the interconnections of *Figure 2-13* are somewhat more complicated than the two- to four-wire hybrid of *Figure 2-14*.

Operation of the Hybrid

The heart of the hybrid system is the multiple winding transformer. By electromagnetic coupling, signals are transferred between windings. Where coupling results in opposing fields, signal cancellation occurs. Thus, two superimposed ac signals can be recovered.

The hybrid is a multiple winding transformer. Actually, as indicated by the diagram in *Figure 2-14*, it usually consists of two interconnected transformers in one physical container. Like any transformer, it uses electromagnetic coupling, i.e., a changing current in winding A produces a changing magnetic field, which in turn induces a changing voltage in winding C on the same core. When the winding C is connected in a circuit, the changing voltage causes a changing current to flow in the circuit. Thus, the changing current pattern or waveform of the input winding (primary) is reproduced in the output winding (secondary) without physical connection between the two circuits. The ratio of the input voltage to the output voltage and the input current to the output current depends on the ratio of the number of turns in the primary winding to the number of turns in the secondary winding. The turns ratio also allows impedance matching between coupled circuits. The impedance in the secondary appears in the secondary circuit as $N^2 \times Z_p$, where N is the ratio of the secondary turns to the primary turns and Z_p is the primary impedance. Impedance matching is important in circuits to reduce power loss and signal reflections.

In *Figure 2-14*, a transmit signal applied to terminals 7–8 is inductively coupled to windings C and D by the current flow through windings A and B. The voltage induced in winding D causes a current to flow through the two-wire circuit connected to terminals 1–2 for further transmission of the signal. This same current also flows through winding F which induces a voltage in winding H.

Figure 2-14.
Hybrid Circuit

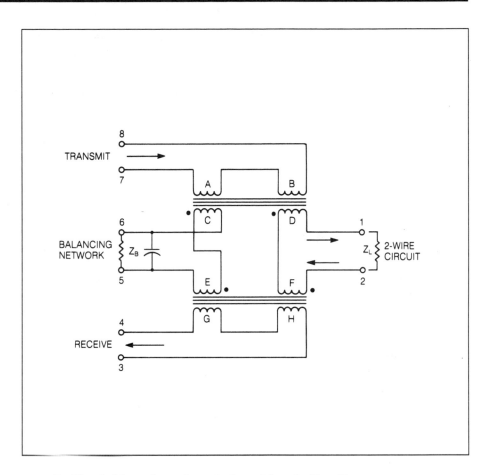

In like fashion, the voltage induced in winding C causes a current to flow through the balancing network and winding E. The impedance Z_B of the balancing network matches line impedance Z_L. Since Z_B equals Z_L, the turns on coil C and D are equal, and the turns on coil E and F are equal, the same current flows in the circuit of C and E and the circuit of D and F. As a result, the same voltage is induced in windings G and H, since they also have turns equal to each other.

Windings C and E have their connections reversed from the connection of windings D and F. Therefore, the voltage induced in winding G is opposite in phase to the voltage induced in winding H and opposes the voltage in winding H. Since they are equal and opposite, they cancel each other. The net result is that the signal from the transmitter that appears at terminals 7 and 8 is transmitted to terminals 1 and 2, but does not appear at terminals 3 and 4 that go to the receiver. (The signal level at terminals 1–2 is only one-half the level input to terminals 7–8 because the other half is dissipated in the balancing network.)

A similar effect occurs when a signal appears at terminals 1 and 2 from the line. The current flowing in windings D and F induces a voltage in windings B and H. The voltage in H causes a current to flow through the circuit of the receiver—winding G, and winding H. The same current that flows through H flows through G. The current flowing in winding G induces a voltage in winding E and causes a current to flow in the balance network. This current flows through winding C and induces a voltage in winding A. The voltage in winding A is 180° out of phase from that induced in winding B by the current flowing in winding D. This again is due to the reversed connection of windings C and E. Therefore, the winding A voltage and the winding B voltage oppose each other and cancel each other. None of the signal that appears at terminals 1 and 2 from the line appears at terminals 7 and 8, the transmitter terminals.

Sidetone

A small amount of sidetone is desirable in the telephone, so the user can hear his/her own voice. A special circuit slightly unbalances the hybrid to introduce a controlled cross-over signal.

For the hybrid used in the central office exchange, the balancing network is adjusted so that no transmit signal appears at the receive terminals and no receive signal appears at the transmit terminals. However, for the induction coil arrangement in the telephone set, the balancing network is intentionally unbalanced slightly so that a small amount of the transmitted signal is also fed to the receiver of the talking phone. This signal is called the *sidetone*.

Sidetone is necessary so that the person can hear his/her own voice from the receiver to determine how loudly to speak. The sidetone must be at the proper level because too much sidetone will cause the person to speak too softly for good reception by the called party. Conversely, too little sidetone will cause the person to speak so loudly that it may sound like a yell at the other end. In the circuit shown in *Figure 2-13*, the values chosen for R2, C2, and C3 and the turns ratio of the windings on the induction coil produce the proper level sidetone for most conditions. Varistor RV2 provides automatic adjustment of the sidetone level for telephone set installations as variations in loop current occur due to the different length of local loops.

THE ELECTRONIC TELEPHONE

A study of *Figure 2-13* reveals an interesting fact; the conventional dial telephone set contains no "active" circuit components (an active electronic element is one which accomplishes a gain or switching action). In other words, the ordinary telephone performs no amplification of the incoming electrical signals. Whatever electrical signal current level is produced at the central office is what is transmitted down the local loop and is what the listener hears. In fact, the signal level at the listener's receiver is weaker because of line loss in the local loop.

Any amplification of the signal occurs in the central office or in the exchanges and repeaters between central offices. This is done to centralize the expensive and complex equipment for maintenance and so the equipment can be shared by many users. As a result, the local loop and central office have become a type of "centralized processing" system; while the telephone set itself has remained a relatively inexpensive "dumb" device that serves only as an interface and input/output device. Said another way, all the "intelligence" is located in the exchanges. However, this is changing and, as will be shown in this book, the telephone set is evolving into a "distributed processing" system, where there is intelligence not only in the central office, but also in the relatively "smart" telephone sets.

Equivalent Circuits

The entire phone system may be better understood in terms of its equivalent circuit, which consists of distributed and lumped values of R, C, and L.

As the discussions in the following chapters show how the telephone set is evolving and will evolve into its position in the distributed system, equivalent circuits of the telephone system will be useful. *Figure 2-15* shows such a circuit. The exchange, line, and telephone set are replaced with equivalent circuits. The series resistance and shunt capacitance of the line are represented by lumped values of resistance R_L and capacitance C_L. The telephone set is represented by the transmitter, receiver, and equivalent impedances of the induction coil Z_{F1} and Z_{F2}. The balancing network made up of R_{B1}, R_{B2}, and C_B is adjusted to match an average line impedance and to provide the correct amount of sidetone.

The transmission equivalent circuit of the telephone is represented by the transmitter being the source of emf, feeding a series-parallel resistance network. Note the net voltage drop across the receiver is about zero.

It is useful to understand the equivalent circuit of the telephone set under the varying conditions of transmission and reception. *Figure 2-16* shows the equivalent circuit when speaking into the transmitter. The voltage generated as a result of speaking is v_T. As it is generated, voltage drops equal to it exist in each of the loops containing Z_L and Z_B. Voltage v_o is the voltage that is transmitted down the telephone line as the output resulting from the speech voltage, v_T. The remainder of the v_T voltage, v_{F1} is dropped across the impedance Z_{F1}. In the balance network loop, v_{ZB} is the drop across the balancing network and v_{F2} is the drop across Z_{F2}. The voltages v_{F1} and v_{F2} are equal and oppose each other; therefore, there is no voltage across the receiver because v_{F1} and v_{F2} cancel each other.

The receiver equivalent circuit of the telephone is represented by the source of emf in series with the load, feeding a series-parallel resistance network. Note that the actual voltage drop across the receiving device is now large enough to drive the device.

Figure 2-17 shows the equivalent circuit when a signal is being received from the line. The signal is represented as a voltage v_r in series with the line impedance, Z_L. The current resulting from v_r produces voltage drops around the network of v_{F1}, v_T, v_{F2}, and v_B. Now v_{F1} and v_{F2} are in the same direction, so that there is a net voltage V_R across the receiver which causes the receiver to produce sound. The voltage across the transmitter, v_T, is just dissipated in the transmitter and causes no reaction.

These equivalent circuits will be useful in analyzing the use of electronic circuits to replace the passive functions present in the conventional telephone set. For example, the purpose of the present configuration of transmitter, hybrid, and receiver is to transfer as much as possible of the incoming audio frequency power to the receiver, and to transmit as much as possible of the speech power into the transmitter to the rest of the network. If the dc supply from the exchange can be used to power a low-cost, low-voltage audio frequency amplifier in the telephone set, then both the electrical circuit configuration and the physical telephone set can change quite drastically.

The conventional hybrid circuit with its audio transformer can be replaced by a smaller and less expensive circuit using resistive bridge techniques, with the amplifier gain compensating for the losses in the bridge. Since the gain of the amplifier can be controlled, modern low distortion microphones can be used and compensation for loop length can be closely controlled. We'll begin the look at such circuits in the next chapter.

**Figure 2-15.
Equivalent Circuit of
Exchange, Telephone
Line, and Telephone
Set**

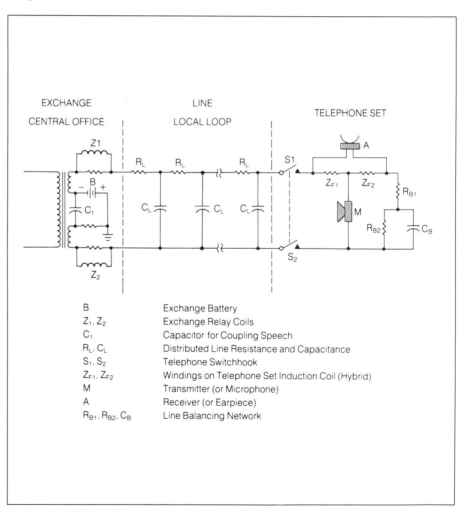

B	Exchange Battery
Z_1, Z_2	Exchange Relay Coils
C_1	Capacitor for Coupling Speech
R_L, C_L	Distributed Line Resistance and Capacitance
S_1, S_2	Telephone Switchhook
Z_{F1}, Z_{F2}	Windings on Telephone Set Induction Coil (Hybrid)
M	Transmitter (or Microphone)
A	Receiver (or Earpiece)
R_{B1}, R_{B2}, C_B	Line Balancing Network

**Figure 2-16.
Telephone Set
Hybrid—Transmission
Equivalent Circuit**

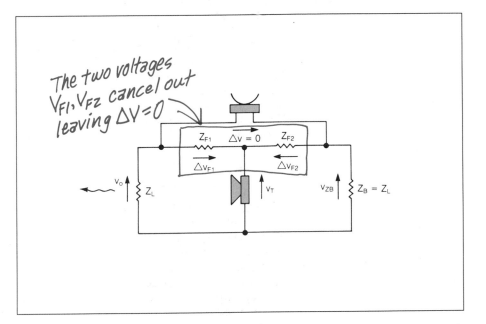

**Figure 2-17.
Telephone Set
Hybrid—Reception
Equivalent Circuit**

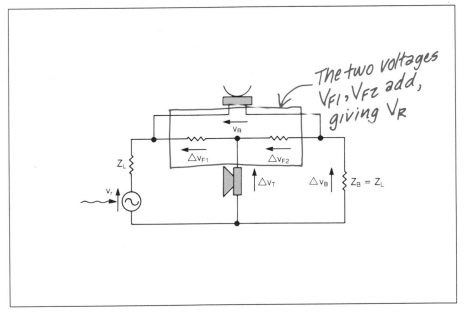

WHAT HAVE WE LEARNED?

1. The conventional telephone set performs the following functions: signals the system that it wants the use of the system, sends the dialed telephone number, transmits speech, receives speech and call progress tones, rings to indicate an incoming call, and signals the system to disconnect when the handset is hung up.

2. The transmitter converts speech into electrical signals when its carbon granule resistance is varied by sound pressure on its diaphragm.

3. The receiver converts electrical signals into sound by applying variations in an electromagnetic field produced by the varying electric current to the receiver diaphragm.

4. The dialing of a called telephone number is equivalent to giving the telephone system an address.

5. Voltages as large as 130 V rms are applied to the telephone set from the line by a ringing voltage generator in the local exchange.

6. Off-hook signal occurs when the handset is off its cradle or hanger and allows current to flow continuously in the local loop.

7. Tone dialing is much faster than pulse dialing and the DTMF tones may be used for low-speed data communications.

8. Muting circuits are used to prevent dial pulse transient voltages from ringing the bell and to protect the speech circuits.

9. The length of the local loop affects the speech signal level so automatic compensation for loop current is provided in the telephone set.

10. A hybrid is a transformer used to interface a four-wire circuit to a two-wire circuit. The induction coil in a telephone set performs this function.

Quiz for Chapter 2

1. The nominal value of the dial pulse break interval in milliseconds in the U.S. is:
   a. 40
   b. 60
   c. 100
   d. 67

2. What type of transmitter is most commonly used in a conventional telephone handset?
   a. carbon
   b. electromagnetic
   c. electret
   d. ceramic

3. Which component in the telephone set has the primary function of compensating for the local loop length?
   a. resistor
   b. varistor
   c. capacitor
   d. induction coil

4. What type of receiver is most commonly used in a conventional telephone handset?
   a. carbon
   b. electromagnetic
   c. electret
   d. ceramic

5. Which component in the telephone set has the primary function of interfacing the handset to the local loop?
   a. resistor
   b. varistor
   c. capacitor
   d. induction coil

6. How many unique tones are used for the 12-key dual-tone multifrequency keypad?
   a. 2
   b. 3
   c. 7
   d. 12

7. Tone dialing takes _____ time than pulse dialing.
   a. less
   b. more
   c. about the same

8. Which of the following are important for proper interface of a DTMF generator to the telephone line?
   a. impedance
   b. tone amplitude
   c. loop current
   d. all of the above

9. The anti-tinkle circuit:
   a. prevents tampering with the telephone.
   b. prevents dial pulsing from ringing the bell.
   c. prevents speech signals from ringing the bell.
   d. all of the above.

10. The sidetone is:
    a. a type of feedback.
    b. determined by the balancing network.
    c. permits the talker to hear his/her own voice.
    d. all of the above.

Electronic Speech Circuits

ABOUT THIS CHAPTER

State-of-the-art electronics concepts and techniques must be designed to be compatible with existing telephone networks. Therefore, major improvements in equipment are made very gradually.

With this chapter, the discussion moves from a description of the telephone system as it presently exists to a discussion of how it is changing and improving as mechanical and conventional electrical devices are replaced by electronic devices, usually in the form of integrated circuits. These devices have most of the required components on the chip, with connections provided for outboard components such as resistors and capacitors which are used to "program" the chip; that is, to set the electrical operating parameters for specific applications. In this way, one design can be used for a variety of similar, yet different requirements. This approach lowers cost and improves performance. In this chapter, the emphasis is on the circuits that provide two-way speech in the telephone set.

But first an important point must be made. Any improvement to any part of the telephone system, whether it be in the telephone set, the switching offices, or the transmission lines, must be not only better, more reliable, less expensive, easier to manufacture, and all the other features associated with progress; but also it must, above all, be compatible with the existing system. If it were possible to start all over again with current technology, the telephone industry undoubtedly would do many things differently. Wiring to subscribers might be in the form of a time division multiplex loop from house to house, rather than a separate pair from each residence to the central office. The bandwidth would certainly be wider; signaling and ringing would be different. But for most systems, the luxury of an entirely new design is not possible. There is not enough capital, manufacturing capacity, or manpower available in the United States to completely rebuild the telephone system, nor is there any reason to want to do so.

The improvements described in this and following chapters will be implemented in the way the telephone companies have managed change for over a hundred years; that is, only small changes which are compatible with the existing network at the interface (the point where it is connected) are made at any given time. Change in the telephone industry is evolutionary, not revolutionary; subtle, rather than sudden.

DC REQUIREMENTS FOR THE LOCAL LOOP

The electronic telephone must be able to operate entirely from the local loop current.

In order to discuss the changes to the circuits in the telephone set, we must consider the portion of the system that we cannot change, but with which the electronic telephone set must interface. This is, of course, the local loop from the central office. It is important that, where possible, the electronic telephone use the loop current from the telephone company central office to provide power for the electronic circuits. Therefore, let's first examine the dc parameters of the local loops seen by the telephone set.

Figure 3-1 shows the typical battery feed circuit arrangements used in central offices to connect to the local loop. *Figure 3-1a* shows a capacitive-coupled circuit and *Figure 3-1b* a transformer-coupled circuit. The configuration at the central office depends on the type of supervisory signaling used on the transmission lines being connected and on how the central office was designed.

In each battery-feed arrangement, L1 and L2 are equal and their sum is the total inductance of the feed circuit in the central office including the sensing relay. The same holds true for the resistance; R1 and R2 are equal and their sum is the total resistance, again including the sensing circuit being used. R_L and C_L are the lumped values for the line resistance and line capacitance. R_{eq} is the equivalent resistance that the telephone set presents to the line for determining the local loop current. V_B is the central office voltage applied to the local loop. *Table 3-1* lists some typical values for the battery feed components in several different systems.

**Table 3-1.
Central Office Battery Feed**

System	Nominal Battery Voltage V_B (volts)	Typical Resistance (Ohms) $R_T = R_1 + R_2$	Current (mA) I_{MAX}	I_{MIN}
1	48 ± 2	400	120	20
2	48 ± 2	800	60	20
3	24 ± 2	400	60	20

For conventional telephones, the carbon transmitter determines the minimum acceptable operating loop current.

For most conventional telephone sets in the United States, the component which determines the minimum operating current in the local loop is the carbon transmitter. It requires 23 mA to operate reliably, but minimum current on the local loop is usually set at 20 mA. (The sensitivity of the line relays in the central office also can determine the minimum local loop current; however, they are designed to operate within the 20 mA limit of *Table 3-1*.) With V_B at 48 volts and R_T equal to 400 ohms, the maximum current that would flow in the local loop is 120 mA if the line were shorted at the central office. This is system 1 of *Table 3-1*. (R_T varies considerably with different countries; e.g., France has 380 to 780 ohms.)

**Figure 3-1.
Typical Central Office
Battery Feed**

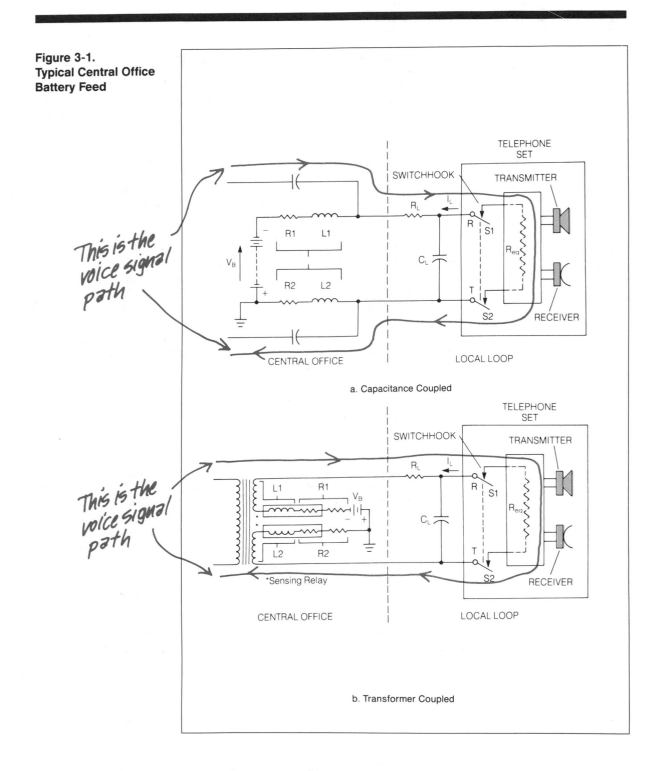

a. Capacitance Coupled

b. Transformer Coupled

The equivalent resistance, R_{eq}, of *Figure 3-1* for the telephone set is typically about 400 ohms (but may be as little as 100 ohms). Using Ohm's law, if 48 volts is divided by 23 mA (the minimum current to operate the carbon transmitter), the maximum local loop resistance is 2,087 ohms. If R_{eq} is 400 ohms and R_T is 400 ohms, the maximum line resistance, R_L, for the local loop is $2,087 - 800 = 1,287$ ohms, or typically 1300 ohms (see *Table 1-7*).

If electronic telephone sets are used with transmitters that are microphones other than the carbon granule type, then the minimum loop current may be much less than 20 mA, say 10 mA. In this case, the limiting component specification is the minimum current required to operate the line relay in the central office.

Under these conditions, the on-hook current drawn by the electronics inside the telephone set must be well below the minimum current required by the line relay, otherwise the line relay would energize and incorrectly indicate an off-hook condition. Obviously, the on-hook current drain also must be held to a minimum so the central office supply will not be overloaded. For these reasons, some local exchanges have specifications for minimum loop current that fall in the range from 6 to 25 mA. This is especially true for local exchanges that may be detecting off-hook loop current with optoelectronic couplers, which are then interfaced directly to digital logic circuits.

On-hook current requirements should be kept to a minimum to prevent the line relay from energizing, and to reduce the subscriber telephone current drain on the central office.

TWO-WAY SPEECH BLOCK DIAGRAM

With the past discussion as background, let's now shift to the main topic of this chapter—the electronic circuits and different devices that replace the transmitter, the receiver, and the transformer-coupled induction coil (or hybrid) of the conventional telephone set.

In order to relate how the electronics fit into the telephone set, the equivalent circuit of *Figure 2-14* for the hybrid circuit is shown again in equivalent form in *Figure 3-2,* modified with the two-way speech electronics. The transmitter now feeds signals to a transmission amplifier marked with the symbol A_T, and the receiver is driven by a receiver amplifier marked with the symbol A_R. The triangle is the common symbol for an electronic amplifier. The amplifiers now couple to the equivalent circuit at the same point in *Figure 3-2* as the transmitter and receiver did in *Figure 2-14*. The balance network of *Figure 2-14* is represented by Z_B in *Figure 3-2*.

Two more circuits that are not shown in *Figure 3-2* are required when electronic circuits are used in the telephone set. These additional circuits, the overvoltage protection and polarity protection circuits, are shown in the block diagram of *Figure 3-3*.

In the electronic telephone, receiver and transmitter amplifiers are used.

**Figure 3-2.
Equivalent Circuit of
an Amplified Speech
Circuit**

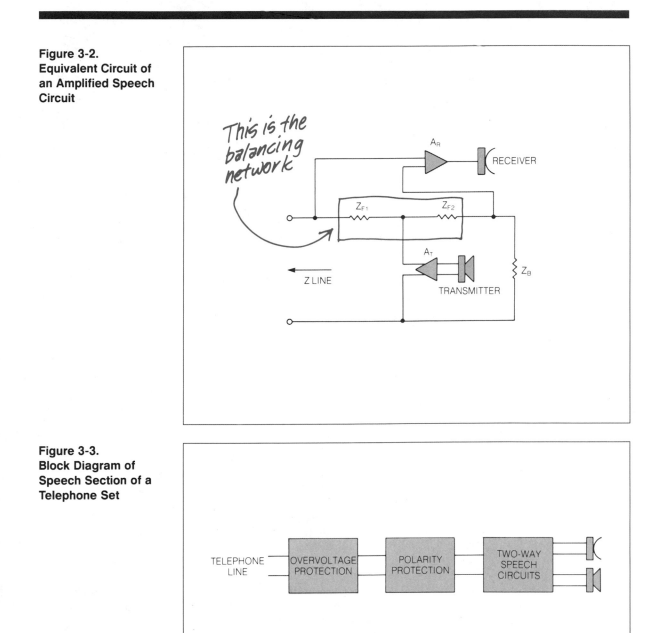

**Figure 3-3.
Block Diagram of
Speech Section of a
Telephone Set**

PROTECTION CIRCUITS

Overvoltage Protection

The electronic circuits are protected from high voltage spikes on the line by zener diodes.

Electronic circuits using small signal transistors and integrated circuits are easily damaged by high voltage transients that may appear on the line due to lightning, switching transients, or power line induction. Therefore, some form of transient and overvoltage protection must be used to protect the circuits in the electronic telephone set. Since the technology used for making integrated circuits cannot provide high enough breakdown voltage, one or more external zener diodes are placed across the line to provide overvoltage protection. When the input voltage exceeds the zener (breakdown) voltage, the zener diode conducts and holds the voltage input to the electronic circuit at the rated zener voltage. If the overvoltage protection is needed at a point where either polarity voltage may appear, the device may be constructed as two zener diodes connected back-to-back, or two physically separate zener diodes may be connected back-to-back across the line.

Polarity Protection

Polarity reversal is prevented by the rectifier bridge.

Polarity of the normal input voltage is critical for electronic circuits since they won't operate if the polarity is reversed, and they may be damaged. The method commonly used to protect against polarity reversal is the rectifier bridge. The output voltage polarity of a rectifier bridge is always the same regardless of input voltage polarity.

Conventional Rectifier Bridge

For line voltage of less than 5 volts, the low-voltage rectifier bridge is used. Its forward voltage drop is low because the conducting transistors are at saturation. The transistors are subject to voltage spike damage, however, and are therefore protected by zener diodes.

A rectifier bridge can be built easily by using four diodes as shown in *Figure 3-4a*. Since the rectifier diodes can withstand high voltage transients, only one zener diode is required on the bridge output to protect the speech circuits.

A conventional full-wave silicon diode rectifier bridge has a forward voltage drop of 1.5 volts. This voltage drop becomes important for electronic telephone sets because they are voltage-operated. Recall the conventional telephone set discussed in Chapter 2 was current operated. Therefore, the minimum line voltage is a much more important specification for the electronic telephone set. Since the speech circuits must have 3.5 volts to operate, the line voltage must be at least 5.0 volts if a silicon diode bridge is used.

Low-Voltage Rectifier Bridge

Some telephone systems may have less than 5.0 volts available. In these cases, the electronic telephone set must use a low-voltage rectifier bridge which has a voltage drop of 0.5 volt or less. This characteristic can be obtained with an active bridge arranged as shown in *Figure 3-4b* so that the voltage drop between input and output is the $V_{CE(sat)}$ of a transistor. $V_{CE(sat)}$ is the voltage drop across the collector-emitter junction when the

**Figure 3-4.
Overvoltage and
Polarity Protection
Circuits**

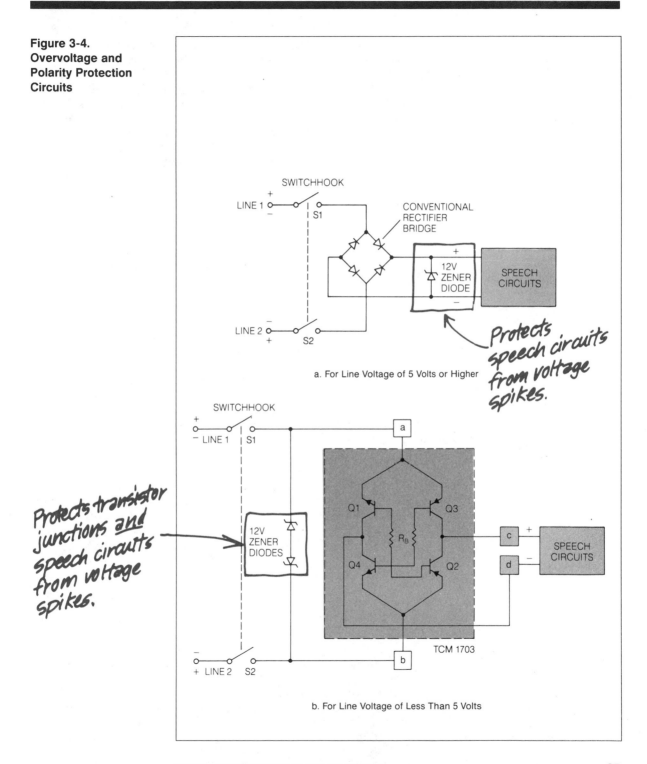

SWITCHHOOK

LINE 1

S1

CONVENTIONAL
RECTIFIER
BRIDGE

12V
ZENER
DIODE

SPEECH
CIRCUITS

LINE 2

S2

Protects speech circuits from voltage spikes.

a. For Line Voltage of 5 Volts or Higher

SWITCHHOOK

LINE 1   S1

a

Q1   Q3

R_B

Q4   Q2

c   +

d   −

SPEECH
CIRCUITS

Protects transistor junctions and speech circuits from voltage spikes.

12V
ZENER
DIODES

TCM 1703

LINE 2   S2

b

b. For Line Voltage of Less Than 5 Volts

junction is in a condition where it is switched On. This operating condition is called saturation. Depending upon the polarity of the input voltage applied to points *a* and *b* either transistor pair Q1–Q2 or Q3–Q4 conducts. If point *a* is positive and point *b* is negative, Q3 and Q4 conduct. Q1 and Q2 conduct when point *b* is positive and point *a* is negative. Thus, the output voltage polarity at points *c* and *d* is the same regardless of the line polarity. The values of resistors R_B are chosen so that the transistors always operate in saturation. The voltage drop across the bridge is the sum of the $V_{CE(sat)}$ of the two "On" transistors.

This bridge circuit also has the desirable characteristics of low shunt current (about one milliampere) and no effect on speech frequencies. However, the maximum line voltage it can handle is about 14 volts. This is sufficient for the normal voltage variations, but high-voltage transients would damage it. Therefore, the transient protection circuit must be placed between the bridge and the line input as shown in *Figure 3-4b*. Since the input voltage may be of either polarity, two zener diodes are connected back-to-back across the line.

THE SPEECH CIRCUIT

A simplified block diagram of a speech circuit suitable for implementation as an integrated circuit is shown in *Figure 3-5*. The Motorola MC34014 is a telephone speech circuit offering built-in transmit, receive, and sidetone circuits, as well as a dc loop interface, regulator, and equalizer circuit. It is connected to the telephone line by a conventional rectifier bridge. The bridge is dynamically equivalent to a small resistance in series with the signal path, and a high resistance in parallel to it. External components are used to adjust transmit, receive, and sidetone gains and frequency responses.

An independent dc loop interface circuit can be optimized with R_5 according to the length of the subscriber line to compensate for variations in loop current. The sidetone level can be adjusted by altering the value of R_2. The MC34014 can be configured to work with either a high impedance (electret), or low impedance (electrodynamic) microphone.

The MC34014 will operate properly with an input voltage (V+) as low as +1.5 Volts. This low voltage operation allows great flexibility in the circuit's operating environment. An interface for dialer circuits (dialing circuits will be covered in detail in the next chapter) is provided to simplify the use of integrated circuit dialers with the speech network. An on-board regulated voltage source and adjustable tone dialing amplifier are also available.

DC LINE INTERFACE

The dc line interface provides a steady voltage source to the circuit and sets the loop current characteristics of the telephone.

The dc line interface controls the dc voltage and current characteristics of the entire speech network depending upon the value of loop current in the subscriber line. Regulator circuitry sets the operating voltage in the integrated circuit and biases the speech circuits. The loop

Figure 3-5.
Motorola MC34014
Basic Speech Circuit

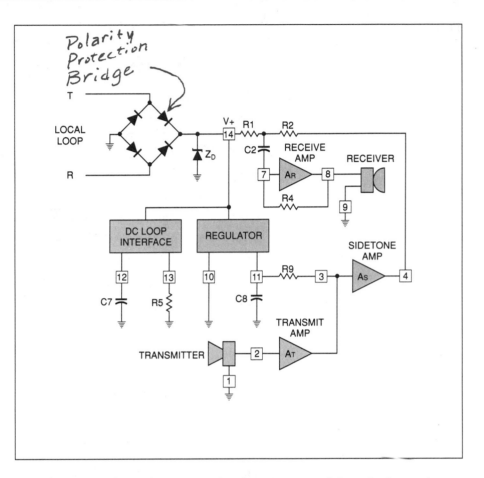

compensation circuit determines the dc resistance of the telephone. An independent voltage regulator (V_{DD}) is included which supplies a steady output voltage for dialing circuits. A block diagram of the dc line interface circuit is shown in *Figure 3-6*.

Loop Compensation Circuit

Loop compensation sets the dc resistance and voltage characteristics of the speech network.

The loop compensation circuit is used to set the dc operating characteristics for the telephone based upon the length of the subscriber loop and the current flowing through it. The dc resistance of the telephone can be set by selecting the appropriate value of R_2. This resistor also helps to dissipate power away from the integrated circuit. A low-pass filter is formed by capacitor C_2. It attenuates any substantial ac voice signals to prevent R_2 from placing an excessive load on speech or tone dialing signals.

Figure 3-6.
DC Line Interface
Block Diagram

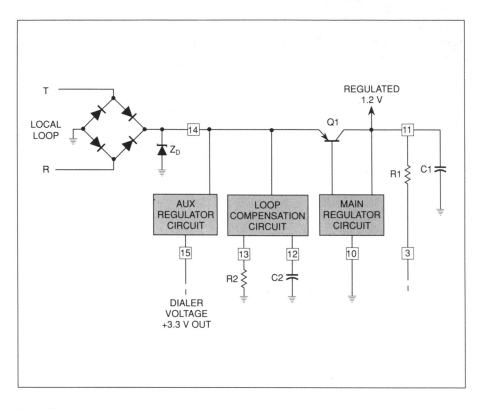

Regulators

A regulator controls the
voltage and current
supplied to a load
circuit.

The main regulator circuit supplies power and bias signals to the
entire speech network. This includes the transmit, receive, sidetone, and
equalization speech amplifiers. It also provides a 1.2-volt signal to bias a
microphone. An external capacitor is often added to stabilize the output
voltage level.

A simple series regulator is shown in *Figure 3-7*. This circuit uses an
NPN transistor in series with the load. A zener diode provides a reference
voltage level. When configured in this way, the transistor behaves much
like an adjustable resistor. The regulator's output voltage equals the zener
voltage minus the voltage drop from the base to the emitter (typically
about 0.6 volt). Load current flows through the transistor from the
collector to the emitter, and then to the load. If the load resistance
decreases, the transistor will automatically become more conductive and
maintain voltage across the load at the higher current, and vice versa.

The Motorola MC34014 is equipped with an auxiliary regulator
circuit which will provide 3.3 volts to any external tone dialing integrated
circuit. This is a shunt-type regulator, similar in principle to the shunt
regulator circuit shown in *Figure 3-8*. The regulating component (Q) is in
parallel, or is shunting, the load.

**Figure 3-7.
Simple Series
Regulator Circuit**

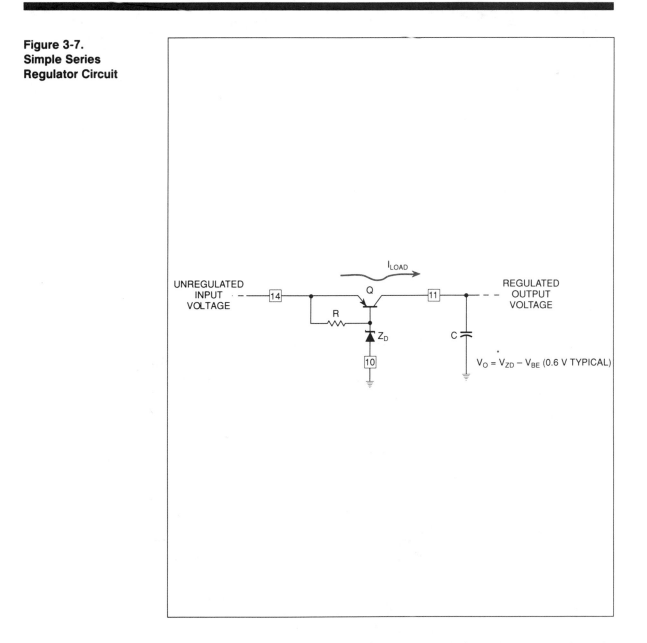

In the shunt regulator, the load voltage will be equal to the value of zener voltage plus the base-emitter voltage across the transistor. A decrease in load will cause a corresponding decrease in V_{BE}, base current, and collector current. Since a smaller load requires less current, I_R decreases along with the voltage across R. This decrease tends to restore the value of the load voltage to its expected level. The reverse is also true for an increase in load.

**Figure 3-8.
Simple Shunt
Regulator Circuit**

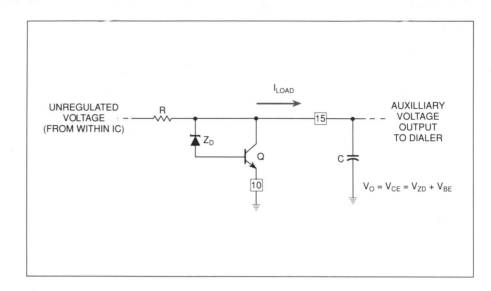

TRANSMITTER SECTION

The transmit circuitry for the MC34014 is shown as a block diagram in *Figure 3-9*. The microphone is biased through a network of external components. A high-impedance electret microphone can be biased as in *Figure 3-10a*. By changing the bias circuit as in *Figure 3-10b*, a low-impedance electrodynamic microphone can be used. Both configurations capacitively couple the microphone signal to the transmit amplifier. A mute circuit is included in the transmit circuit to automatically turn the microphone off whenever dialing is performed.

Transmit Amplifier

The properly biased microphone signal enters the transmit amplifier, A_T at pin 2. A_T is an operational amplifier (or op-amp) configured as an inverting amplifier. All bias voltages and currents necessary to power the amplifier are provided internally based upon the output of the regulator in the dc loop interface. *Figure 3-11* is a typical diagram of an inverting amplifier.

An inverting amplifier offers several useful characteristics in the telephone circuit. First, the amplifier gain can be fixed at a constant value regardless of the loop current level. This relationship is defined with the ratio of the feedback resistor (R_F) to the input resistor (R_I). Gain is given as follows:

$$A = -5\ R_F/R_I$$

**Figure 3-9.
Transmit Block
Diagram**

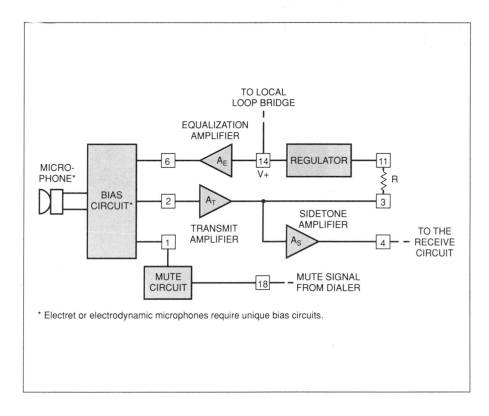

**Figure 3-9.
Transmit Block
Diagram**

The negative sign in the formula indicates that the output signal will be 180 degrees out of phase with the input signal. The transmit amplifier in the Motorola MC34014 has a gain of 20 (about 26 dB).

A second useful characteristic of the inverting amplifier is that the input impedance of the amplifier can be determined by the choice of R_I. Due to the characteristics of the circuit, input impedance will essentially equal R_I, or:

$$Z_{IN} = R_I$$

The input impedance of the transmit amplifier is typically about 10 kΩ.

A bias voltage at the positive input of the amplifier sets up a center voltage on which the microphone signal will vary. The transmit amplifier will then amplify the positive and negative portions of the microphone signal in order to modulate the loop current.

Output voltage from the transmit amplifier is converted to current through resistor R_1 in *Figure 3-6*. This current is fed through the regulator's pass transistor, in turn modulating the loop current. It is this action which places transmitted speech on the local loop. A portion of the

Figure 3-10.
Microphone Biasing

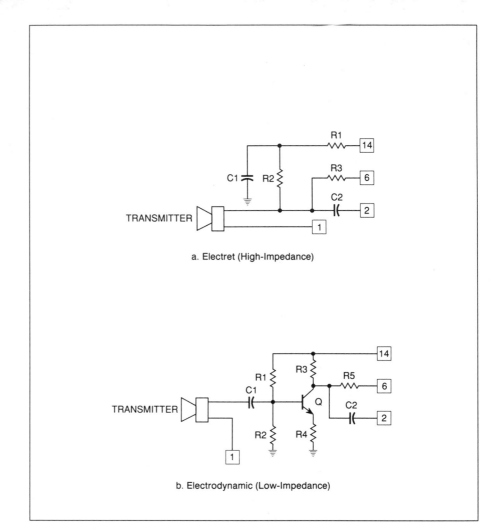

a. Electret (High-Impedance)

b. Electrodynamic (Low-Impedance)

transmitted signal is introduced to a sidetone amplifier, as well as an equalization amplifier.

Sidetone Amplifier

The sidetone amplifier, A_s in *Figure 3-12*, provides a portion of the transmitted signal to the receiver. The person speaking can then hear his or her voice in the receiver and judge the proper volume at which to speak. The sidetone amplifier performs its function by inverting the transmitted signal and supplying a portion of the signal to the receive circuit as shown in *Figure 3-12*. Resistors R_1 and R_2 select the desired amount of sidetone cancellation. The presence of C_1 provides a measure of phase shift to the signal to adjust for the phase shift introduced by the local loop.

Figure 3-11.
An Inverting Amplifier

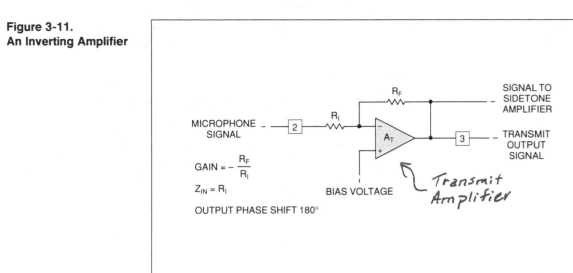

Figure 3-12.
Sidetone Amplifier

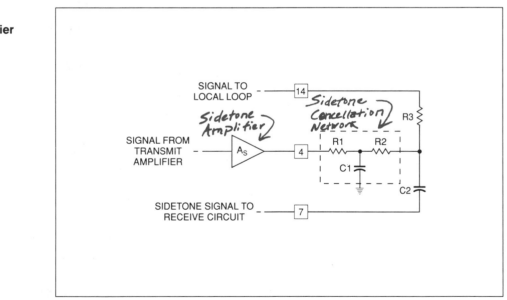

Unlike the transmit amplifier, sidetone gain is directly proportional to loop current. For low loop currents (typically found on long loops), sidetone gain is about 0.17 (or −15 dB). High loop currents (encountered on short loops) will reduce this gain to 0.09 (or −21 dB). As a result, signals on a long loop will be amplified more than signals on a short loop. This tends to keep sidetone amplitude constant regardless of loop length.

The sidetone amplifier will compensate for differences in loop current.

Note also that the sidetone gain is always less than 1. Since only a small portion of the transmitted signal is required back at the receiver, it is necessary to attenuate the speech before it is sent to the receiver.

Equalization Amplifier

The equalization amplifier, A_E, adjusts the bias level to the microphone circuit to compensate the variations in the loop current. This improves the modulation characteristics of the microphone for all loop lengths. Equalization gain, like the sidetone gain, is directly related to loop current. At a high loop current, the gain is about 0.75 (or −2.5 dB). Low loop currents will yield a gain of only about 0.25 (or −12 dB). The equalization amplifier does not invert the signal, so it will be in phase with the transmitted signal. *Figure 3-13* shows a simple diagram of the equalization amplifier.

**Figure 3-13.
Equalization Amplifier**

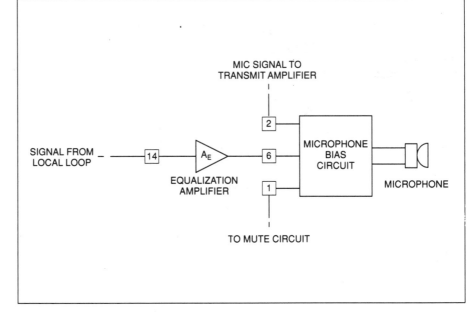

The gain of the equalization amplifier is also less than 1. This limits the effect of the equalization at the microphone bias network. In actual operation, the effects of equalization will have little noticeable effect upon the quality of the transmitted speech.

RECEIVER SECTION

Receiver circuitry for the Motorola MC34014 is shown in the block diagram of *Figure 3-14*. The section is composed of the receiver amplifier itself, as well as a mute circuit and the output of the sidetone amplifier covered in the preceding section. Passive components are used to select the parameters of the circuit's operation.

**Figure 3-14.
Receive Block
Diagram**

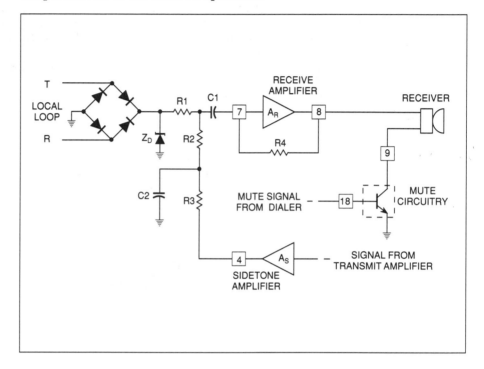

Receiver Amplifier

The local loop enters the receive circuit at pin 7 where it is amplified by the receiver amplifier, A_R, and coupled to the receiving element. Just about any voice channel signal on the local loop will be amplified. A small sidetone signal from the sidetone amplifier is mixed with the receive signal to supply sidetone in the telephone handset.

Gain in the receive amplifier is composed from a variety of factors. Its primary component is the ratio of R_4 to R_1 (the gain of an inverting amplifier). However, the introduction of sidetone brings many other factors into play—gain of the equalization, sidetone, and transmit amplifiers, along with effects from other discrete components must be considered. The specific formula for receiver gain in the MC34014 can be found in the *Motorola Telecommunications Device Data Book*.

Mute

The mute circuit prevents loud noises on the receiver during dialing.

A mute circuit is also added to the receiver network of *Figure 3-14*. A mute works in conjunction with the speech network's dialer interface to shut down (or "mute") the receiver during dialing operations. This prevents loud tones or "clicks" from reaching the receiver. The mute control signal is generated by the external dialing circuit.

In practice, the mute circuit is little more than an open collector transistor in line with the ground side of the receiving element as shown in *Figure 3-15*. During normal speech, the mute signal is a logic 1. This will turn the transistor on and provide a low impedance path to ground. The receiver will work normally. When dialing, the mute signal will become a logic 0. The switching transistor will turn off and break the flow of receiver current. This effectively disables the receiver.

Figure 3-15.
Simple Mute Circuit

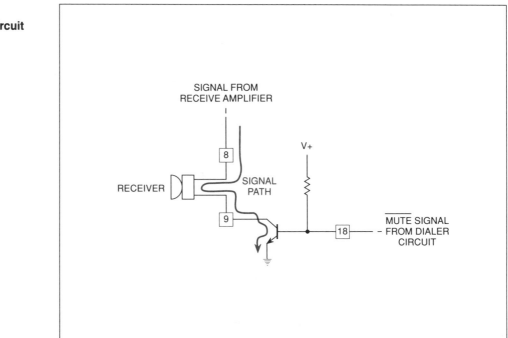

Dial Interface

The Motorola MC34014 contains a tone amplifier and interface circuit designed to connect dialing signals directly to the loop. *Figure 3-16* is a diagram of the dial interface. Pin 17 is a mode select input which switches the mute logic between tone and pulse dialing modes. In this way, the circuitry can provide the appropriate muting signals for tone or pulse dialing. Mode selection must also be connected to the respective dialing circuit.

The mute signal from the dialer is sent to pin 18. It is this input which supplies the logic control signal that mutes the receiver. Tone dialing signals can be connected directly to pin 16. These signals will be amplified directly by the dial amplifier, A_D, and sent along to the local loop (more on dialers in the following chapter).

Figure 3-16.
Dial Interface

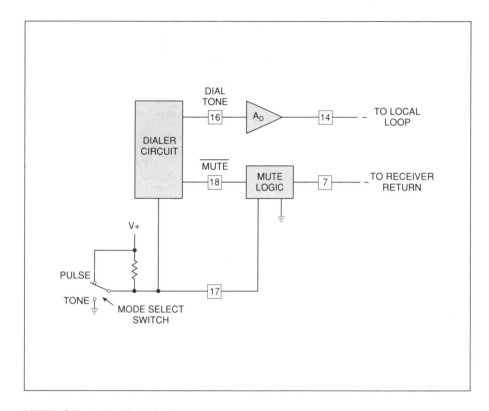

NETWORK IMPEDANCE

The most substantial problem encountered with older telephone designs was that their performance would vary depending on the length of the local loop. A telephone on an unusually long loop would be hard to hear, while a telephone on a short loop could be annoyingly loud. These old telephones would often be optimized for the particular local loop, but a telephone adjusted for a long loop would be far too loud if used on a short loop, and vice versa. A process of equalization was needed to automatically compensate the impedance of the speech network for variations in the local loop. It was not until the introduction of varistors that passive network equalization became feasible. The integrated circuit speech network in the MC34014 emulates this performance with active components.

The ac impedance of the speech network is dependent on the transmitter amplifier, the equalization amplifier, and three external resistors. *Figure 3-17* shows the circuit relationships for determining speech network impedance, Z_{NET}. The relationship is given by the formula:

$$Z_{NET} = \frac{1 + (R_2/R_1) \times (R_3)}{20 \times A_E \times (R_2/R_1)}$$

where,

A_E is the gain of the equalization amplifier.

The gain of the equalization amplifier can be anywhere from 0.25 to 0.75 depending on the loop current. This variation will allow the network impedance to vary by a factor of 3.

**Figure 3-17.
AC Impedance
Equivalent Circuit**

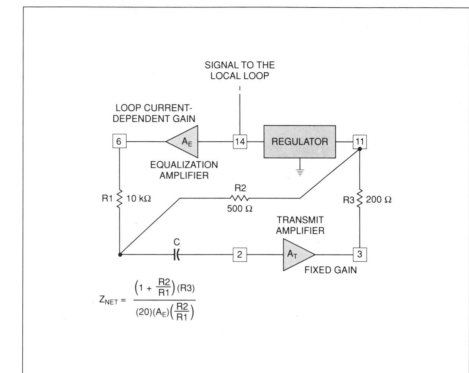

The range of impedances for the speech network can be calculated by inserting the known resistances from *Figure 3-17* into the formula, then performing two separate calculations—one at minimum equalization gain (A_E) and one at maximum equalization gain. Inserting values for minimum gain encountered on a long loop:

$$Z_{NET} = \frac{(1 + (500/1000)) \times 200}{20 \times 0.25 \times (500/1000)}$$

reducing this first equation:

$Z_{NET} = 840$ ohms

Inserting the values for maximum gain found on a short loop:

$$Z_{NET} = \frac{(1 + (500/1000)) \times 200}{20 \times 0.75 \times (500/1000)}$$

reducing this second equation:

$Z_{NET} = 280$ ohms

Equalization will allow the speech network to work well on local loops of any length.

This exercise demonstrates the flexibility of the integrated speech network to match itself to the local loop for optimum performance. For the resistor values specified, the impedance of the MC34014 will vary from a low of 280 ohms when connected to a short loop, to 840 ohms when on a long loop.

LINE BALANCING

The discussion of the telephone set would not be complete without considering line balancing. We shall discuss this topic from a general view, considering both electrical and acoustic factors effecting a classical passive telephone instrument.

In the overall operation of a telephone, several factors affect the transfer of energy between the twisted pair, the receiver, and the transmitter amplifiers.

The entire telephone set, including the handset, the amplifiers, and the hybrid, is shown in *Figure 3-18*. A signal a from the microphone results in the a-b portion being delivered to the telephone line amplified by a factor K based on the gain of A_T and the ratio of decoupling of the hybrid. The imbalance of the hybrid results in a fraction of this signal, b, being fed through the receiver amplifier, A_R, to the receiver. This is the sidetone as discussed in Chapter 2.

The electrical signal to the receiver is converted into sound and fed to the user's ear. However, a fraction, b_x, of that acoustical signal will be fed back to the microphone via the air and possibly the facial bones of the user. This acoustical feedback can be positive and add to the original signal, a, to cause howling if the level of feedback is high enough. The acoustical feedback can be reduced by the shape of the handset, the mounting of the receiver and microphone inside it, and the material from which it is made.

The electrical signal balancing depends on the hybrid network and the value of line balancing impedance Z_B. The latter values are determined by the telephone company after a great deal of statistical survey on the lines used by the telephone system and on the spread of parameters of the production components that go into the telephone set.

Figure 3-18.
Signal Paths

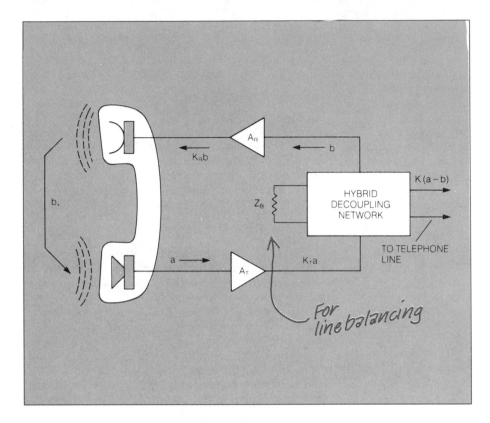

TELEPHONE SET VOLUME COMPENSATION

The electronic tele-
phone maintains a con-
stant output level over
a wide range of loop
currents, by automati-
cally adjusting the
transmitter and re-
ceiver amplifier gains.

Recall in the discussion of the conventional telephone set that there
were varistors (components whose resistance changes as the current
through them changes) in the hybrid circuit that adjusted their resistance
so that the volume of the sound in the telephone system from transmitter
to receiver stayed about the same even though the length of the local loop
increased. *Figure 3-19* illustrates this feature for the common Bell Type
500 telephone set. Note that the horizontal axis is either local loop
resistance R_L in ohms (top of chart) or line length in thousands of feet
(bottom of chart). The vertical axis is the response in decibels.

The curve of the relative response of the Bell Type 500 telephone set
receiver shows that the relative volume level delivered to the receiver
stays within ±1 dB for a local loop length of 0 feet ($R_L = 0$) to a length of
14,000 feet ($R_L = 1180$ ohms). As a comparison, an older telephone set,
the Type 302, had a 5-dB loss in volume when the local loop was 12,000
feet long. Similarly, another curve in *Figure 3-19* shows the transmitting
response. As explained for the two-way speech integrated circuit, a signal
from the regulator based on the current in the local loop adjusts the gain
of the transmitter amplifier and the receiver amplifier of the electronic

**Figure 3-19.
Conventional
Telephone Set
Relative Volume
Levels vs Line Length**
*(Courtesy Bell
Laboratories)*

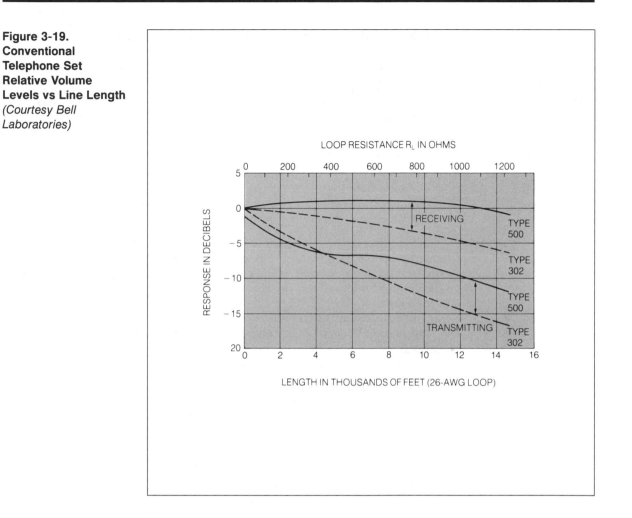

telephone to provide the same or better compensation than that of the Bell
Type 500 set.

AVAILABILITY AND CHARACTERISTICS OF IC SPEECH NETWORKS

Motorola is just one manufacturer of telephone integrated circuits.
Integrated circuit speech networks and other telephone component
products are available from a wide variety of IC manufacturers such as
Texas Instruments, Exar, Oki, and Samsung. Each manufacturer offers a
range of different products, each with its own features, characteristics, and
options. Full performance details for any component such as the Motorola
MC34014 can be obtained from the respective manufacturer.

At first glance, the Motorola MC34014 data sheet may appear
unwieldy, but there are several main characteristics that are critical to the

speech network's performance. *Table 3-2* lists some of the most important points.

**Table 3-2.
Motorola MC34014
Characteristics**

Characteristic	Min	Max
DC Line Voltage	+1.5 Volts	+15.0 Volts
Transmit Input Impedance		10 kΩ
Transmit Gain		26 dB
Tone Input Resistance	1.25 kΩ	
Tone Gain	3.2 dB	
Sidetone Gain	−21 dB	−15 dB
Equalization Gain	−12 dB	−2.5 dB

Note that the receive gain is not given as a specification. This is because receive gain is a combined function of transmit, sidetone, and equalization gains, as well as the effects of external components. It is this relationship which allows the receiver volume to compensate the changes in loop current.

To emphasize the flexibility of the IC speech network, *Table 3-3* describes the effects of each component shown in the application of *Figure 3-20*.

**Table 3-3.
External Component
Functions**

Component	Function
R1, R4	Determines a portion of receive gain
R2, R3	Controls sidetone cancellation
R5	Determines dc resistance of speech network
R6, R8	Microphone biasing resistors
R7	Controls tone input amplitude
C1	Controls sidetone phase shift
C2	Couples signals to receive amplifier
C3	Couples signals to receiver
C4	Stabilizes receive amplifier
C5	Couples microphone signals to speech network
C6	Stabilizes dc loop voltage
C7	Low-pass filter for R5
C8	Stabilizes regulated voltage output
C9	Stabilizes regulated dialer voltage

Figure 3-20.
MC34014 Application
Circuit. *(Courtesy Motorola, Inc.)*

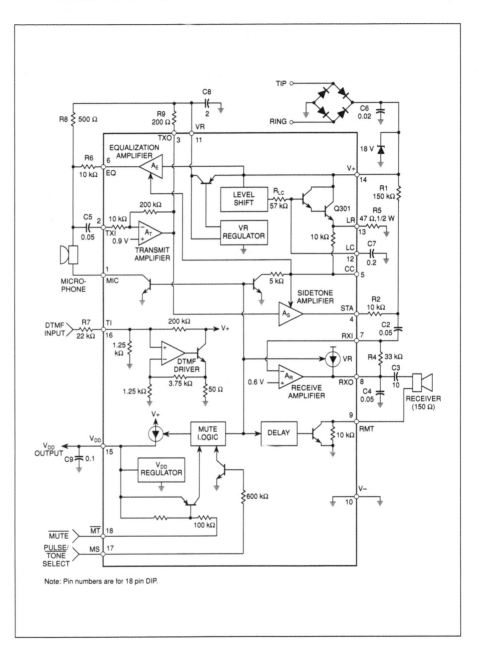

WHAT HAVE WE LEARNED?

1. Changes to the telephone set must be made in a manner that maintains the proper characteristics for the line interface.
2. The coupling of two-way speech circuits into the telephone set is similar to the conventional hybrid circuit except a resistive bridge replaces the induction coil.
3. Two-way speech circuits are made with integrated circuits because they offer high performance, low power consumption, small size, are lightweight and have high reliability.
4. Two-way speech circuits must operate from very low voltage (below 4 V) if powered from the telephone line.
5. Overvoltage protection must be provided to prevent damage to electronic circuits by voltage variations and spikes on the line.
6. The two-way speech circuits must operate with either polarity voltage on the telephone line. A rectifier bridge is used to provide a constant polarity voltage.
7. Two-way speech circuits must draw very low leakage current from the telephone line in the on-hook condition.
8. Circuits for voltage and current regulation are contained in the two-way speech integrated circuit.
9. Controlled gain is provided by the two-way speech amplifiers in the electronic telephone set to compensate for the line length. This is in contrast to the conventional telephone set which compensates for line length by using varistors.
10. Either an electrodynamic or an electret microphone may be used with the proper choice of input bias circuitry.

Quiz for Chapter 3

1. The equipment used in the telephone network changes:
   a. rapidly, to keep up with new technology.
   b. almost never, because it is doing the job.
   c. slowly, to balance new technology against cost.

2. The telephone set can change:
   a. very little, because it is connected to a network that changes slowly.
   b. as long as the interface to the subscriber loop remains the same.
   c. only in cosmetics, such as color, shape, style, etc.

3. The carbon transmitter:
   a. is a comparatively recent electronic development.
   b. was invented by Thomas A. Watson, Bell's assistant.
   c. determines the minimum loop current.

4. On-hook current must be kept low so that the:
   a. line relays in the central office will not mistake it for off-hook current.
   b. comparatively small wires in the cable will not overheat.
   c. ringer will not ring incorrectly.
   d. carbon microphone will not be damaged.

5. Power to operate the telephone and the subscriber loop is usually provided by:
   a. the telephone company, and is low voltage ac.
   b. the subscriber, through the use of batteries in the set.
   c. the central office, and is called battery feed because batteries are sometimes used.

6. Overvoltage protection is:
   a. needed in the telephone set because of the high ringing voltage.
   b. is nearly always incorporated on the IC chip.
   c. is primarily a central office function.
   d. needed because of transients from dial pulsing, lightning, induction, or short circuits.

7. A special rectifier bridge is used in electronic telephones because:
   a. the ac power must be converted to dc to simulate a battery.
   b. it is necessary to shunt the speech frequencies and keep them off of the loop.
   c. the voltage drop across conventional bridges leaves too little voltage to operate the set.

8. Voltage and current regulation
   a. is provided by the central office.
   b. is needed to ensure proper IC operation.
   c. can be performed by circuits built into the IC.
   d. b and c above.

9. The equalization amplifier
   a. compensates the microphone for changes in loop current.
   b. controls the frequency of dialing tones.
   c. mutes the microphone.

10. Electret and electrodynamic microphones are directly interchangeable devices in the speech network.
    a. True
    b. False

Electronic Dialing and Ringing Circuits

ABOUT THIS CHAPTER

In this chapter, the discussion centers on replacing the conventional telephone set components that perform the functions of dialing and ringing with electronic circuits that do these functions. One might say that electronics has made the mechanical rotary dial obsolete. Solid-state integrated circuits are available that generate dial pulses or tones when a telephone number is entered by depressing keys on a keypad. Electronic memories can retain one or more phone numbers so that the touch of only one key on a keypad sends the entire number down the line. Likewise, electronic ringers are replacing the bell to signal an incoming call. Single-tone and multitone electronic ringers are available.

DIAL PULSE GENERATION USING INTEGRATED CIRCUITS

Old electromechanical dialers have been replaced by integrated circuits. However, their output signal characteristics must remain the same to stay compatible with existing central office equipment.

Dial pulsing is the most commonly used method in the world for sending addresses (telephone numbers) to the central office. Consequently, the mechanical rotary dial with its gears, cams, contacts, and rotary speed governor was one of the first targets for the substitution of electronics. The rotary dial itself is being replaced with a push-button keypad and the pulse generating mechanism is being replaced with an integrated circuit called an impulse dialer. Several versions of this circuit are made by a number of semiconductor manufacturers.

In order to be compatible with electromechanical central offices, the electronic impulse dialer must send dial pulses at the 10 pulse per second rate. Since a person can push the buttons faster than the pulses can be sent, an electronic memory capable of holding from 17 to 20 digits is included in the circuit to hold the number while the dial pulses are transmitted. The number remains in this memory for several hours or until another number is entered.

Redial

Using integrated circuits provides the redial feature and meets the requirement for very low power.

With the memory function available, it was relatively easy to incorporate a redial function. By pushing only one button, the last number entered will be redialed. This feature is helpful when continually trying to reach a busy number. Initially, the redial feature met with some resistance from the telephone companies because of fears of clogging the network since redialing could be done so rapidly and frequently. Also, they were concerned about the maintenance of batteries used to power the electronic circuits in the telephone set. (Remember the station battery at each telephone was abandoned in favor of a common battery at the central office exchange very early in the 1900s because of cost and maintenance.) The development of complementary metal-oxide-silicon (CMOS) semiconductor technology, which produces integrated circuits with very low power consumption, has solved the latter problem. The current drain of these circuits is only a few microamperes so that it is possible to operate the electronics from the dc power available from the telephone line. However, if a battery of the type commonly used to power calculators or other consumer electronic goods were used to provide only the very small current required by the number memory, the battery would last for several years.

CIRCUIT POWER AND VOLTAGE TRANSIENTS

Before proceeding with the description of the electronic pulse dialer, there are two general subjects that need to be covered in a bit more detail. One is circuit power and the other is voltage transients.

Circuit Power

All telephone electronic circuits must be powered through a rectifier bridge to prevent polarity reversing damage.

Figure 4-1 shows a number of ways that power is provided to electronic circuits inside the telephone set that are powered from the line. The ringer circuit must be supplied from the line ahead of the switchhook so that the circuit can be energized by the ringer signal even though the telephone handset is on-hook. In this case, a diode rectifier bridge must precede the circuits (if it is not provided in the ringer circuit) to protect against line voltage polarity reversal.

Other circuitry in the telephone must also receive power from the telephone line. These additional circuits, however, must be powered after the switchhook assembly. For this reason, an independent diode bridge may be used to supply voltage and current to any device not already containing internal diode bridges.

Although *Figure 4-1* shows standard diode bridges, low-voltage diode bridges may be used to supply power to integrated circuits that must operate on very-low-voltage lines. A well-designed telephone circuit needs no more than two bridge circuits (only one if the ringer has a built-in bridge). Notice the back-to-back zener diodes following each bridge. As discussed in Chapter 3, these diodes provide integrated circuits with a measure of overvoltage or transient protection.

Voltage Transients

Contact making and breaking will cause inductive circuit elements to generate high voltage pulses. Therefore, high voltage transistors and overvoltage protection devices are used.

When the handset is on-hook, the switchhook is open and the full exchange battery voltage (usually 48-V dc) appears at the tip and ring terminals of the telephone set. When the handset is lifted, the switchhook closes. If there is any contact bounce (momentary closures and openings) and the circuit in the telephone set is inductive, relatively high voltage pulses called transients will be generated. Dial pulsing also produces these high voltage transients.

**Figure 4-1.
Circuit Power for Line
Driven ICs**

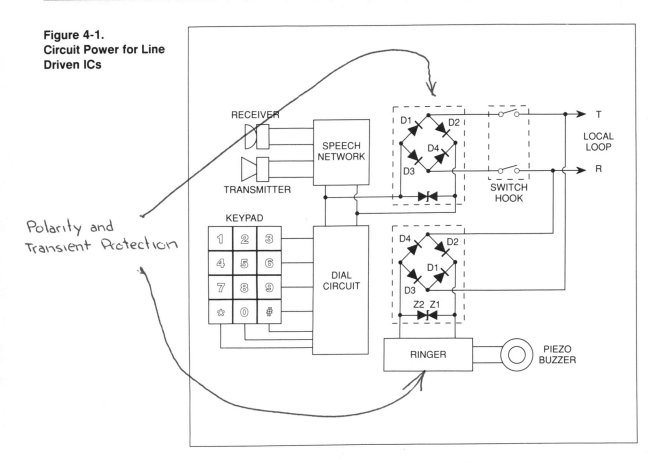

Polarity and
Transient Protection

The magnitude of the induced voltage will be increased by higher current values applied to a larger inductance. Also, the faster the current is applied or removed, the greater will be the induced voltage.

Figure 4-2 shows an equivalent circuit of such a case just as the switch in the circuit has opened. A current I had been flowing. Due to the magnetic field stored in the inductance L, which collapses when the current stops, a voltage is generated across the inductance that appears across the opening switch contacts. This voltage can be calculated by the following equation:

$$V = \frac{L\Delta I}{\Delta t}$$

where
    L is the inductance (in henrys) in circuit,
    ΔI is the change in current (in amperes) that occurred,
    Δt is the time (in seconds) it took to make the current change.

Figure 4-2.
Equivalent Circuit for
Producing Voltage
Transients

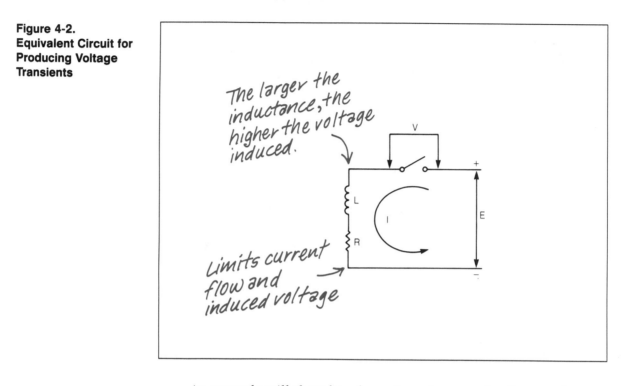

An example will show how large the voltage can be. Suppose L equals 1 henry of inductance, that 10 milliamperes (10×10^{-3} amperes) of current were flowing when the switch opened, and that it took 10 microseconds (10×10^{-6} seconds) to break the circuit. By substituting in the previous equation, the voltage can be calculated.

$$V = \frac{1 \times 10 \times 10^{-3}}{10 \times 10^{-6}}$$
$$= 1 \times 10^{3}$$
$$= 1,000 \; volts$$

The voltage generated is 1,000 volts. Integrated circuits and many small transistors cannot withstand this much voltage and will be permanently damaged. Thus, the electronic phone must have high-voltage transistors in circuits that interface to the line and must have circuits to suppress or absorb these transient voltages. The overvoltage protection, speech-muting, and ringer-muting (anti-tinkle) circuits perform this function.

Overvoltage protection was discussed in Chapter 3. Recall that either single-polarity or double-polarity zener diodes are used depending on the expected polarity of the voltage at the point of protection. The speech-muting and ringer-muting circuits are discussed in the following sections.

PULSE DIALER

The electronic pulse dialer must perform two functions—loop disconnect dialing and receiver muting.

Recall that the pulsing switch on the rotary dial of the conventional telephone set is in series with the induction coil (hybrid) and interrupts the loop current a number of times equal to the number dialed; one time for the number 1, five times for the number 5, etc., at 10 pulses per second. This is called loop disconnect dialing. The impulse dialer electronic circuit that replaces the mechanical dial must do the same function—interrupt the loop current at 10 pulses per second.

It also must do another function. Recall that the conventional mechanical dial has additional switches to short out the handset receiver so that the dial pulses were not heard by the caller. This is called muting. The impulse dialer also must mute the receiver or, as more generally stated, mute the speech circuits. Generally, there are two ways that these functions are accomplished.

Dial Circuits in Parallel with Speech Network

The first way of accomplishing the functions is shown in the equivalent circuit of *Figure 4-3.* Here the dial circuits, represented by switch S3, are in parallel with the speech network. This previously has been called the hybrid network; but more recent terminology for this equivalent is speech network. The speech network is muted by switch S4 under control of the pulse dialer. Fundamentally, here's how it works.

For circuits where the speech network and pulse dialer are in parallel, the pulse dialer must provide an output that pulses the loop current to the line, but is interrupted to the speech network.

When the handset is taken off-hook, S4 is closed and S3 is open. The loop current flows through the speech network in order to signal the central office that a subscriber wants service. When the caller starts to dial and depresses the first key, the pulse dialer circuit closes S3 and opens S4 so that the loop current now flows through the dial pulse circuit and at the same time the speech network is disabled so the receiver is muted. As the circuit dials, it does the equivalent of opening and closing S3 electronically to interrupt the current in the loop the required number of times. When dialing is finished, S3 remains open and S4 closes to maintain the loop current.

Dial Circuits in Series with Speech Network

For circuits where the speech network and pulse dialer are in series, the pulse dialer must provide an output that pulses the line and the speech network, but the current flow to the receiver only is interrupted.

The pulse dialer also may accomplish the functions of interrupting the loop current and muting as shown in *Figure 4-4.* In this case, the circuits to do the current interruption, represented equivalently by S3, are in series with the speech network. Just as in *Figure 4-3,* the pulse dialer does the equivalent of opening and closing S3 electronically. However, unlike *Figure 4-3,* where no current flowed through the speech network while dialing was going on, now the interrupted loop current is flowing through the speech network. Therefore, muting must be accomplished differently.

Electronically, the pulse dialer does the equivalent of opening S4 in the receiver circuit. Thus, even though the current is pulsing through the hybrid network, the noise does not affect the receiver because its circuit is open. Let's now look at a typical integrated circuit pulse dialer.

Figure 4-3.
Dial Circuits in Parallel
with Speech Networks

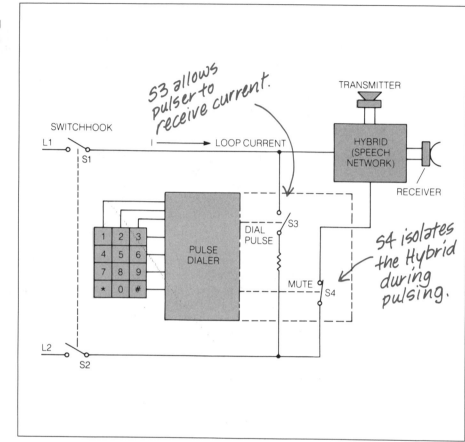

Primary Functions

Figure 4-5 details the major operating sections contained in a typical pulse dialing integrated circuit. Depending upon the particular device, the key decoding logic will accept either a 4-bit binary word corresponding to the selected key, or it may accept the key as a series of row and column lines. For the integrated circuit used in these examples, a binary input will be used.

When a key is detected and verified, the binary code of the number is stored in internal memory. Certain dialer ICs may store anywhere from 16 to 20 digits per phone number. Additional control circuitry in the memory directs the location at which each digit is stored or recalled. Numbers stored in memory will be held until a new number is entered, or until power is removed from the dialer.

Figure 4-4.
Dial Circuits in Series
with Speech Networks

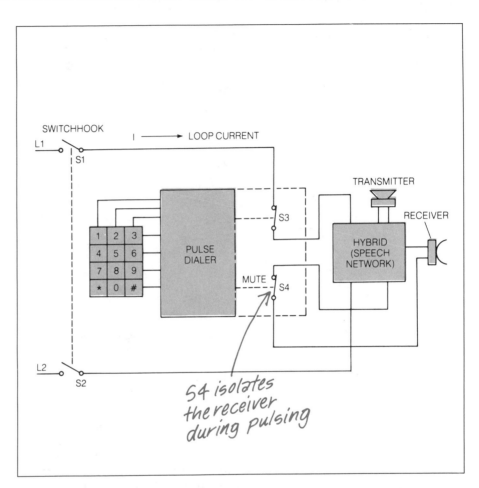

An oscillator circuit is provided to supply a stable clock frequency for all internal dialer circuitry. It is this oscillator that coordinates the timing and sequences of control logic in the dialer. Crystals, resistors, capacitors, or inductors can be used to set the frequency of the oscillator.

The output circuit is directed by control and timing logic and is comprised of a pulse output and a mute signal which are both connected to the speech network. A discrete, medium-power, high-voltage transistor at the IC output uses dial pulses to actually interrupt loop current. Additional discrete components may be needed to employ the mute signal depending on compatibility of the particular speech network. An IC speech network such as the Motorola MC34014 covered in Chapter 3 can use mute signals directly. A speech network using discrete components, such as an older 500-type telephone, would need some sort of mute interface.

Control and timing logic typically has a number of extra input options to control other facets of dialer operation. The three most

Figure 4-5.
Pulse Dialer Integrated
Circuit

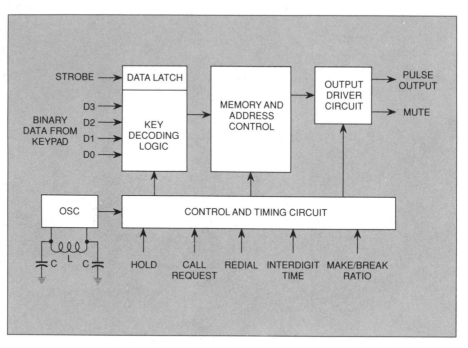

common control inputs are; On-Hook/Off-Hook, Interdigit Time, and Make-Break Ratio.

On-Hook/Off-Hook, sometimes referred to as Call Request, is a signal that enables dialers to sense a switchhook closure and loop current flow. The dialer will not function if this signal is in an On-Hook condition. Interdigit Time, or IDT, controls the delay between subsequent digit pulse sequences. Typical delay values can range from 200 to 1000 milliseconds when dialing at 10 pps, and 100 to 500 milliseconds when dialing at 20 pps. A make/break ratio can be set at either 67% low/33% high, or 61% low/39% high depending on the logic state of the make/break ratio (MBR) input line. The clock frequency directly effects the dial rate, so the 10 or 20 pps dialing rate may be selected by changing the oscillator components.

Two extra functions that are often available are; Hold and Redial. Hold is a signal which disables outpulsing then the current digit is completed. When this line is released, outpulsing will resume normally. This function is very useful in extending the Interdigit Time beyond the normal conventions beyond the IDT line. Redial, as the name implies, will cause the last digit entered into memory to be repeated at the output.

IC Impulse Dialer Application

Dedicated impulse dialers are continuing to loose popularity as more and more central office facilities are being equipped to handle tone-dialing signals. Although central offices will remain compatible with pulse-dial signals into the foreseeable future, the speed and convenience of tone dialing, combined with the onset of automated sales and information services, will eventually render pulse-dial telephones extinct.

One of the few integrated circuit pulse dialers still on the market today is the MC14408 manufactured by Motorola. It is fabricated using low-power CMOS technology and it incorporates a wide range of functions. Since the MC14408 requires a binary input, a keypad-to-binary encoder (such as a Motorola MC14419) must be used to interface the keypad to the pulse dialer.

Originally, the intent of the MC14408 was to provide a direct replacement for the electromechanical rotary dialer mechanism in conventional telephones as shown in *Figure 4-6. Figure 4-7* shows the same conventional telephone circuit with the addition of the integrated pulse dialer.

Figure 4-6.
Standard Rotary Pulse
Telephone *(Courtesy Motorola, Inc.)*

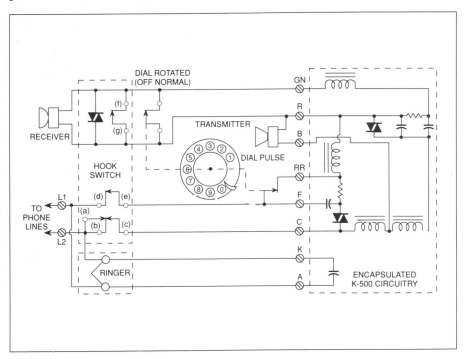

As soon as the telephone goes off hook, the Call Request (CRQ) signal becomes logic low. This resets the MC14408 internal circuitry to either accept a new number or execute a redial function. When the pulser executes a number, the Dial Rotating Output (DRO) becomes logic high. This shuts off the transistor providing a current path to the receiver and effectively mutes it. DRO will stay high while digit pulses are being sent. Dial pulses (*Figure 4-8*) are sent through the Outpulsing (OPL) pin which switches a diode-transistor network on and off to interrupt loop current. The central office interprets these current pulses as digits.

**Figure 4-7.
Modified Pulse
Telephone** (*Courtesy
Motorola, Inc.*)

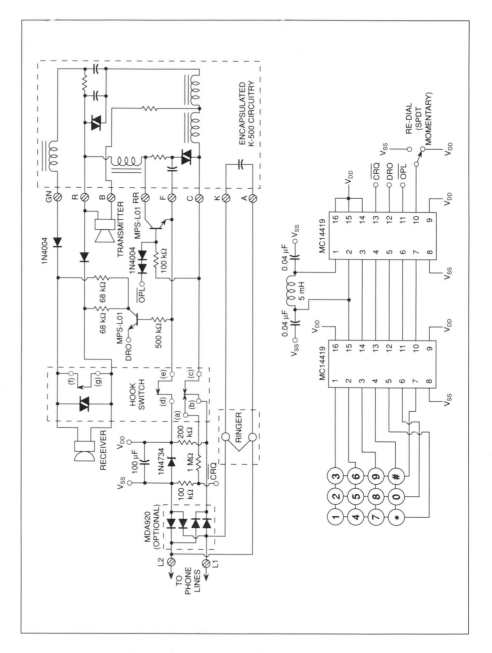

Power (V_{dd}) for the integrated circuits is channeled through an MDA920 polarity protection bridge rectifier and developed across a 1N4734 zener diode and 100-µF capacitor. Ground potential in the circuit is called V_{ss}. The charge stored in the 100-µF capacitor will be enough to retain the dialer memory for several hours.

Figure 4-8.
Pulse Dialer Timing

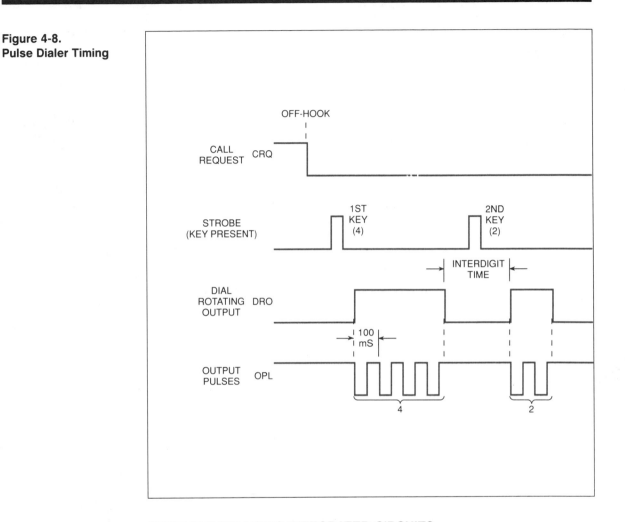

DTMF DIALING USING INTEGRATED CIRCUITS

To generate DTMF tones electronically, keypad closures are digitally converted into combinations of low and high frequency sine waves, which are mixed in pairs and amplified, to drive the speech network.

Dialing also may be accomplished by sending dual tones onto the line as discussed for the conventional phone. Integrated circuits have been designed to provide this function. The conventional way of accomplishing this is shown in *Figure 4-9a*. Unlike the conventional way, in which a low-frequency and high-frequency sine wave oscillator feed the speech network, the integrated circuit DTMF generator (*Figure 4-9b*) has a counter and decoder that counts pulses from a crystal-controlled oscillator and provides output codes that correspond to the low-frequency tone required and the high-frequency tone required. Each of the two outputs from the counter feed into its own digital-to-analog (D/A) converter. The D/A converter, as the name implies, converts the digital code out of the counter to a sine-wave tone.

Figure 4-9.
DTMF Generators

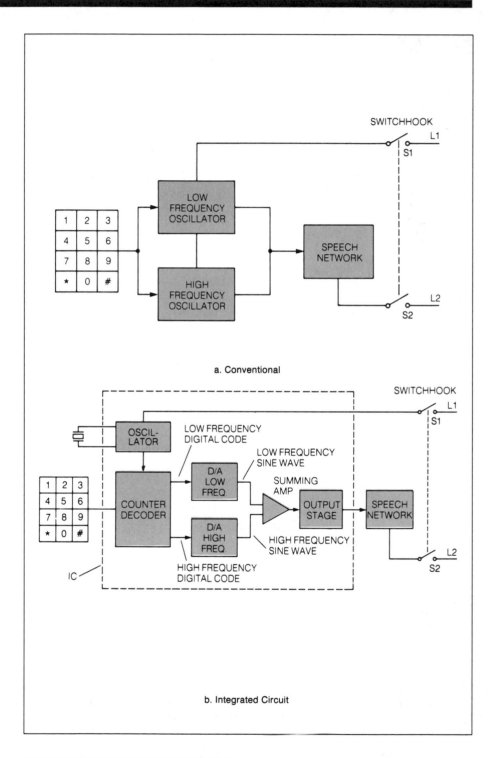

The low-frequency tone and the high-frequency tone are summed in an operational amplifier and are fed to the speech network as a combined signal by the output stage.

An Example IC

Activation of the tone output circuits in the integrated circuit DTMF generator begins with the caller pressing a key on the keypad. The keypad contacts may be arranged as shown in *Figure 4-10a*. The top schematic is a representation of a standard DTMF keypad (like a double-pole single-throw switch) where a separate set of two contacts give the row and the column of the key being pushed. An alternative and somewhat less expensive arrangement is shown in the bottom schematic which is a so-called Class-A or single-pole single-throw configuration, with only one set of contacts for each row-column intersection. Some ICs are designed to accept both types; others accept only one type. In either case, the closed keypad contacts give an indication of the key being depressed. In some cases, the closed contacts may apply a supply voltage on the output lines for as long as the key is held down. In other cases, a ground or the common side of the power supply may be provided. The keypad may be arranged to provide only a pulse, as is the case for a keypad that is used with a scanning sequence. The output waveforms are shown in *Figure 4-10b*.

**Figure 4-10.
Keypad Contact Types
and Output
Waveforms**

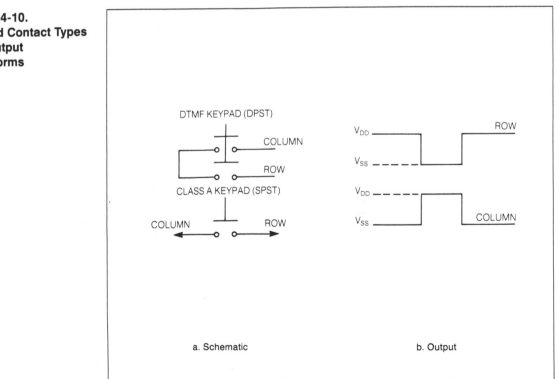

a. Schematic   b. Output

Figure 4-11 shows more detail of a typical DTMF integrated circuit, the TCM5087 (MK5087). This circuit fits into a whole family of devices from TCM5087 to TCM5094. In the diagram, it can be seen that the decoder circuits (labeled = 1) and the set of OR and AND gates in the center have the function of controlling the output of the dividers labeled CTRDIV K. The control is such that if only one key is pushed, two outputs will be generated; one from the row dividers, and one from the column dividers. These two digital outputs are converted to analog waves and summed in the summing amplifier. Note, however, that it is possible to depress two keys on the keypad simultaneously. If the two keys depressed are in the same row or in the same column, and the Single-Tone Enable (Pin 15 on the example circuit) is a 1, only one tone, corresponding to the row or column in which the two keys are pushed, is generated. This feature is implemented to allow testing of the keypad and tone generator circuits. Operation of two keys in diagonal positions causes the circuit to provide no output at all (both sides of the decoder indicate ''not equal to one'').

One other function of interest is shown by the logic element labeled ≥ 1 at the bottom of the high frequency decoder. This gate provides an output at standard logic voltage levels (labeled Mute Out, Pin 10), and an output through a transistor switch (Pin 2), for muting of the speech circuit. The Mute Out is used to mute the receiver during dialing so the caller does not hear the full level of the dialed tones. These outputs are active whenever one or more keys are pushed.

Frequency synthesis in the circuit is done using a single input frequency of 3.579545 MHz, derived from an external crystal oscillator (using an inexpensive color TV crystal) connected to pins 7 and 8. This input frequency is divided in the blocks labeled CTRDIV K by eight different constants to produce eight possible output frequencies, two of which are selected each time a key is pressed on the associated keypad. The two tones are then added and amplified to produce the dual-tone output.

Output Waveforms

The output waveforms are not pure sine waves, but are stairstepped because of the digital-to-analog conversion. Typical stairstep approximation of the row and column sinusoidal output are shown in *Figure 4-12*. The sinusoidal waveforms are generated with a typical distortion of under 7%. The typical harmonic and intermodulation distortion content of the resulting two-tone sum is 30-dB down when referenced to the strongest column tone fundamental.

**Figure 4-11.
Block Diagram of
TCM5087**

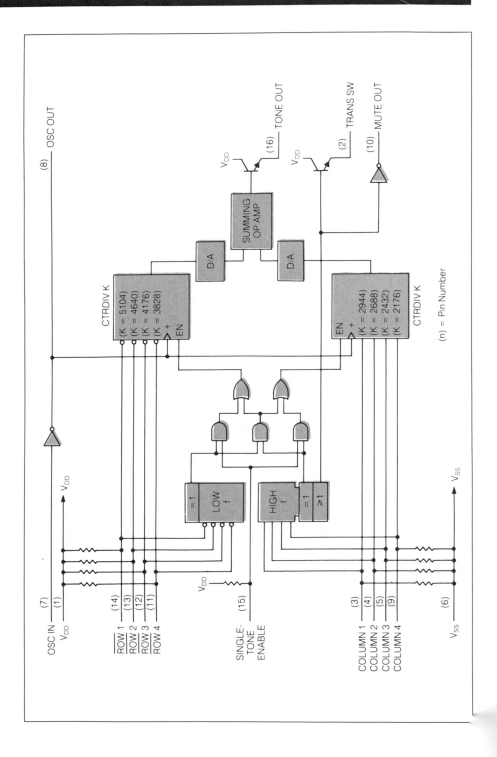

Figure 4-12.
DTMF Waveforms

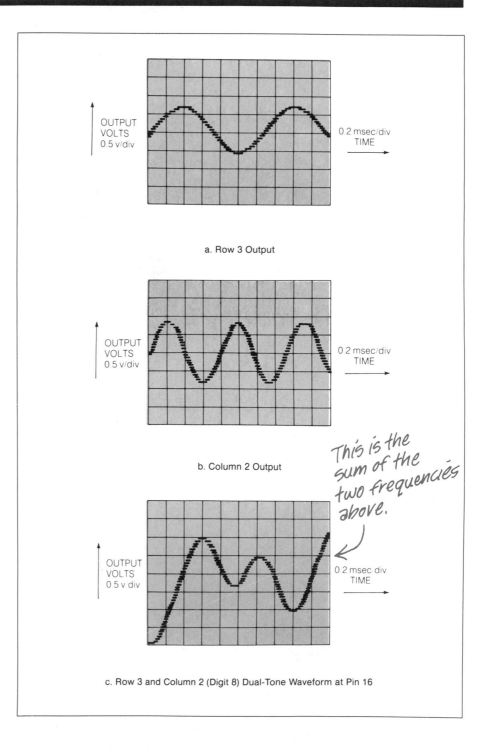

a. Row 3 Output

b. Column 2 Output

This is the sum of the two frequencies above.

c. Row 3 and Column 2 (Digit 8) Dual-Tone Waveform at Pin 16

Actual Application

A TCM5087 is shown applied in a telephone set in *Figure 4-13*. The tone output feeds through the speech network to the line. A diode bridge feeds power to the IC and Z1 protects against transients. The capacitor C1 maintains and filters the voltage across the IC to eliminate noise and supply variations. A mute signal controls Q1 and mutes the transmitter of the speech network. The XMIT output on pin 2 and the ST1 input on pin 15 were described in the discussion of *Figure 4-11*.

**Figure 4-13.
TCM5087 in Telephone
Set**

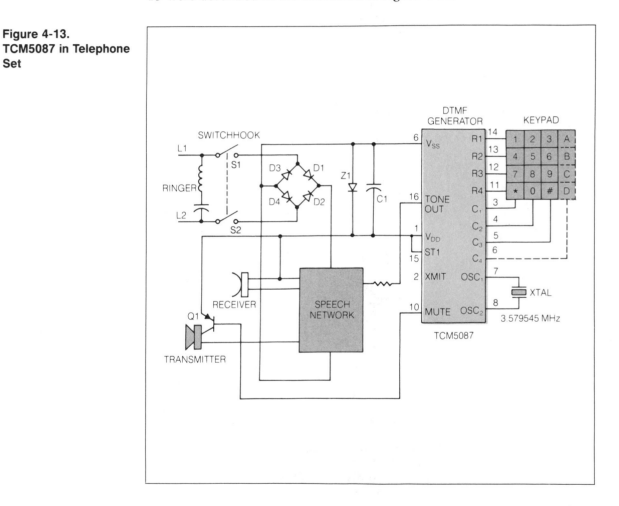

COMBINED DIALERS

Advances in integrated circuit fabrication technology have made it practical to combine the circuitry for a pulse dialer and DTMF dialer into one device. Dialing modes can be selected with a single switch. These combined dialers can also offer last number redial, as well as memory storage for 10 or more telephone numbers.

Actual Components

Motorola manufactures a series of combination dialing circuits; the MC145412, MC145413, and MC145512. *Figure 4-14* is a block diagram for the MC145412/13/512 family. Each dialer interfaces directly with either 3 × 4 or 4 × 4 keypads. A single input pin will select between DTMF, 10 pps, or 20 pps dialing. An internal memory can hold up to 10 complete telephone numbers, each up to 18 digits long—this includes last number redial. Finally, the dialers operate on line power as low as 1.7 volts. This makes them very closely compatible with the MC34014 integrated speech network covered in Chapter 3.

**Figure 4-14.
Combined Dialer Block
Diagram.** *(Courtesy
Motorola, Inc.)*

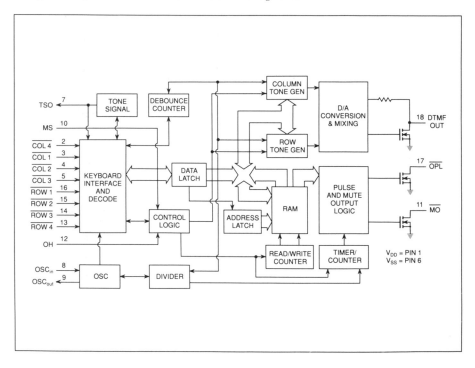

Figure 4-15 demonstrates an actual application of the MC145412. A standard 3 × 4 keypad is used to select desired digits. Keypad signals are connected to row and column inputs. A ground condition at the off-hook input will enable the dialer. While on-hook, numbers may be entered into memory without dialing signals being generated. A 3.58 MHz colorburst crystal provides a stable time base for operation.

**Figure 4-15.
Combined Dialer
Application**

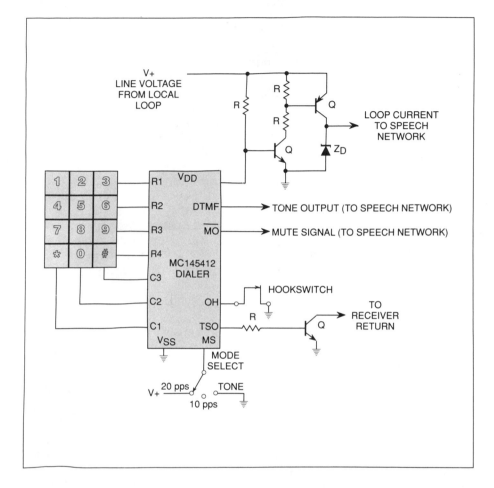

Dialing mode select can be logic high for 20 pps, open for 10 pps, or ground for DTMF signalling. DTMF tones are output to the DTMF OUT pin when the dialer is in the tone mode, and the pulse output (OPL) will be at high-impedance. In pulse dialing mode, the DTMF OUT pin is at high-impedance while pulses are delivered to the OPL pin. Mute is connected directly to the speech network. It will be logic low during pulsing or tone output, otherwise it will be logic high.

Make/break ratios are not field adjustable in the MC145412. Devices are mask programmed at the factory. Both the MC145412/13 offer a typical make/break ratio of 40/60. The MC145512 has a make/break ratio of 32/68.

Memory access and last number redial are accomplished easily by the integrated dialing circuits. For last number redial, the "*" and "0" buttons must be pressed. To retrieve a stored number, "*" and the corresponding number (1 through 9) must be pressed. A complete explanation of dialing key sequences is available in the MC145412/13/ 512 datasheet.

ELECTRONIC RINGER

Electronic ringers have controllable pitch and volume. Also, they are smaller and lighter than conventional bells.

Ringing is the way the called party is signaled that a call is waiting. Conventionally, this has been done with an electromagnetic bell. An electronic ringer, like the conventional bell, must be efficient, cheap, sturdy, and reliable. Unlike bells, however, electronic ringers can provide extra features. They can sound varying pitches so that each of several telephones in an office can be identified. They can sound with increasing volume the longer the phone rings.

An important advantage provided by electronic ringers is their smaller size and lighter weight. New types of telephone sets are possible when the set does not need to contain a large heavy bell that is loud enough to be heard all over the house. Since many homes now have extension phones in one or more locations, this loudness is not needed. Component layout can be flexible so that the circuit can be fitted into unusually shaped housings or can be combined with other electronics onto a main circuit board. Electronic ringers may be either single tone or multitone. Recall that to cause the ringing, the central office places a large amplitude ac signal on the local loop.

Single-Tone Ringer

The single-tone ringer is a self-resonant oscillator, and transducer, which are powered by the ac ringing current.

The single-tone ringer has a fixed-frequency self-resonant oscillator which is turned on and off by alternate half cycles of the ac ringing voltage. *Figure 4-16* shows two single-tone ringer circuits. An electromagnetic transducer is used in *Figure 4-16a* and a piezoelectric "sound disc" type transducer is used in *Figure 4-16b*.

Voltage Regulation

The voltage regulator of the single-tone ringer causes the oscillator period of operation to be determined by the amplitude and frequency of the ringer signal.

The voltage regulation portion of both circuits in *Figure 4-16* is the same. On the negative half cycles of the ac input, D1 conducts and the power is dissipated in R1. The small voltage drop across D1 assures that D2 blocks current flow so no power is supplied to Q1 in the oscillator section; thus, there is no audio output. On the positive half cycles, D2 conducts when the voltage rises above the zener voltage of D3 and the oscillator operates for the duration of the half cycle. Therefore, the time of oscillation is determined by the amplitude and frequency of the ringer signal placed on the line.

Anti-Tinkle Circuit

D3 and C2 form the anti-tinkle circuit to prevent operation of the ringer by dialing pulses. The zener voltage of D3 establishes a threshold voltage that the dial pulse voltage must exceed before it can activate the ringer and C2 is a filter to absorb any transients that exceed the threshold.

**Figure 4-16.
Single-Tone Electronic
Ringer Circuits**

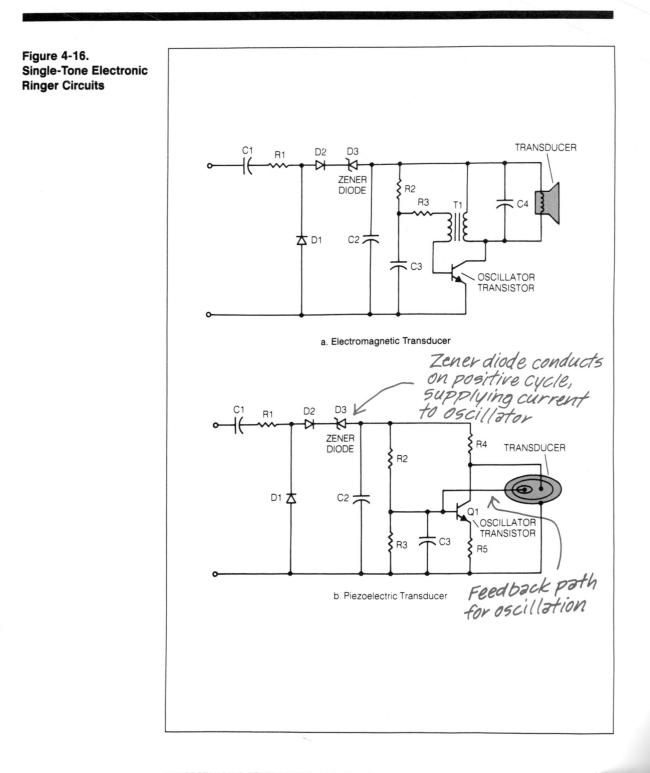

a. Electromagnetic Transducer

*Zener diode conducts
on positive cycle,
supplying current
to oscillator*

b. Piezoelectric Transducer

*Feedback path
for oscillation*

Tone Generation and Output

The magnetic transducer in *Figure 4-16a* typically is a moving-coil or moving-armature transducer similar to those used for telephone receivers described in Chapter 2. Feedback for oscillation is provided by transformer coupling of T1.

The piezoelectric transducer is a thin brass disc to which a disc of piezoceramic material is glued with an epoxy cement (*Figure 4-17*). The upper surface of the ceramic is silver plated to form one side of the electrical connection. Feedback for oscillation of the circuit of *Figure 4-16b* is obtained from a small area of the disc which is isolated from the main area by cutting or etching away the silver around the feedback element. When the circuit oscillates, because of its piezoelectric properties, the disc flexes as shown in *Figure 4-17b* and produces sound. For greatest efficiency, both types of transducers must be mounted in an acoustically resonant enclosure which normally is made as part of the telephone case.

The piezoelectric transducer is a dual-purpose component. It produces sound and also provides the feedback path for the oscillator.

Figure 4-17. Piezoceramic Transducer

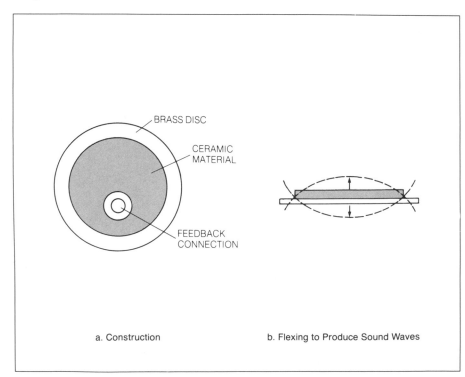

BRASS DISC

CERAMIC MATERIAL

FEEDBACK CONNECTION

a. Construction

b. Flexing to Produce Sound Waves

Disadvantages

The single-tone ringer can be made with few components at low cost, but the output tone is usually around 2 to 3 kHz. At these frequencies, two human factors problems arise;

1. As people get older, their hearing sensitivity at these frequencies is reduced.
2. The ability of people to locate the source of a sound at 2 to 3 kHz is poor.

This latter problem increases the difficulty of determining which phone is ringing in an office with several phones of the same type on different lines. Both of these problems can be solved by using a multitone ringer at a lower frequency.

Multitone Ringer

For the multitone ringer, the only function of the ac ringing voltage is to energize it.

Multitone ringers are necessarily more complex electronically than single-tone types, but with modern integrated circuits, their component count can be even lower than that of the single-tone circuits. The output signal of a multitone ringer is produced by switching between two or more frequencies at a rate determined by the tone ringer circuitry, whereas the ac ringing voltage frequency determines the switching rate of the single-tone ringer. Thus, for the multitone ringer, the ac ringing voltage's only function is to supply power to the unit.

The block diagram of *Figure 4-18* shows the circuit functions typically required in a multitone ringer. The ac ringing voltage is rectified to obtain the dc supply power which is then held constant by the voltage regulator. The overvoltage protection circuitry is the same type that has been discussed previously.

**Figure 4-18.
Multitone Ringer Block
Diagram**

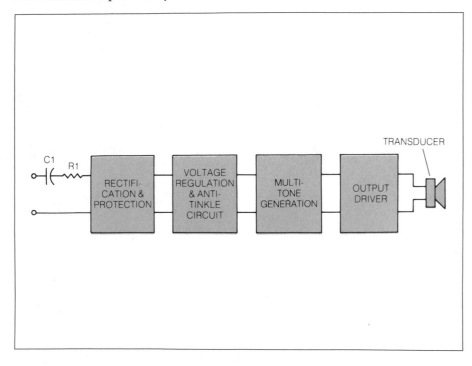

Voltage Regulation

Voltage regulation in the multitone ringer permits normal ringer operation over a wide range of voltages.

The ac voltage at the input to the ringer can vary over a wide range depending on how far the telephone is from the exchange. Voltage regulation is needed so that the voltage applied to the tone generation circuit is independent of the loop length. The regulator must work over a range of input voltage from about 10 to 90 V rms. Usually the voltage regulator will provide an output higher than the minimum operating voltage (say 25 to 40 V). At lower input voltages, reduced performance is tolerated.

Anti-Tinkle Circuits

The anti-tinkle circuit must distinguish between the signal generated during dial pulsing and the ac ringing voltage. The pulses applied to the ringer during dialing are short-duration, high-voltage pulses of up to 200 V at a 10-Hz rate as shown in *Figure 4-19a.* The ac ringing voltage input is roughly sinusoidal at between 16 to 60 Hz.

Figure 4-19.
Anti-Tinkle Circuit
Waveforms

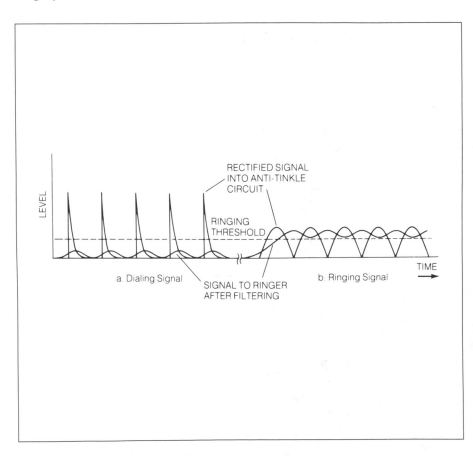

To keep the telephone from ringing during dialing, one type of anti-tinkle circuitry suppresses the spiked dial pulses going to the ringer using frequency selective circuitry and threshold detection.

One way to distinguish between the signals is to make the circuits frequency selective. The dialing transients can be suppressed by making the tone ringer respond only to the higher frequency of the ac ringing signal. The disadvantage of this method is that components for filters at these low frequencies are physically large. Another way is to reduce the sensitivity of the ringer to short pulses with a low-pass filter followed by a threshold detector. The threshold is set so that after filtering the rectified waveform, the transient signal from the dial pulses never rises above the threshold, but the ac ringing signal does, as shown in *Figure 4-19b*.

In practice, this is not as easy to implement as it appears since the threshold setting also limits the level of ringing signal that will produce ringing. On long loops with more than one telephone set connected, the available ringing voltage can be as little as 10 V rms. Therefore, some compromise may have to be made between anti-tinkle performance and ringer sensitivity.

Tone Generation

Two-tone ringing may be accomplished using two oscillators enabled by a frequency control, or by using one oscillator and frequency division.

Several different techniques for tone generation are used in multitone ringers. The most common circuit produces two tones, but there are some three-tone integrated circuits available. Techniques also exist for generating more complex musical sounds.

One of the simplest schemes for generating a two-tone ringing signal is shown in *Figure 4-20a*. This system uses two oscillators—one set to operate at a low frequency (e.g., 10 to 20 Hz) and the other at a higher frequency (e.g., 440 to 480 Hz). A frequency control signal switches the output between the two frequencies during ringing. This approach has the disadvantage of requiring two sets of frequency determining components which increases size and cost.

An alternative two-tone circuit uses a master oscillator to generate a high frequency from which the audio tones and the switching signal are generated by division as shown in *Figure 4-20b*. The master oscillator frequency f is divided first by x or y to give an output of frequency f/x or f/y Hz. This tone is then divided by Z to give a switching control rate of around 10 to 20 Hz which switches the main dividers between X and Y. For most ringers, an external resistor, as shown in *Figure 4-20b*, is used to adjust the frequency of a noncrystal-controlled oscillator. However, to improve frequency stability, the master oscillator in some circuits is crystal controlled as shown by the dashed lines. When crystal controlled, the pitch can be varied only by changing the crystal.

**Figure 4-20.
Multitone Ringers**

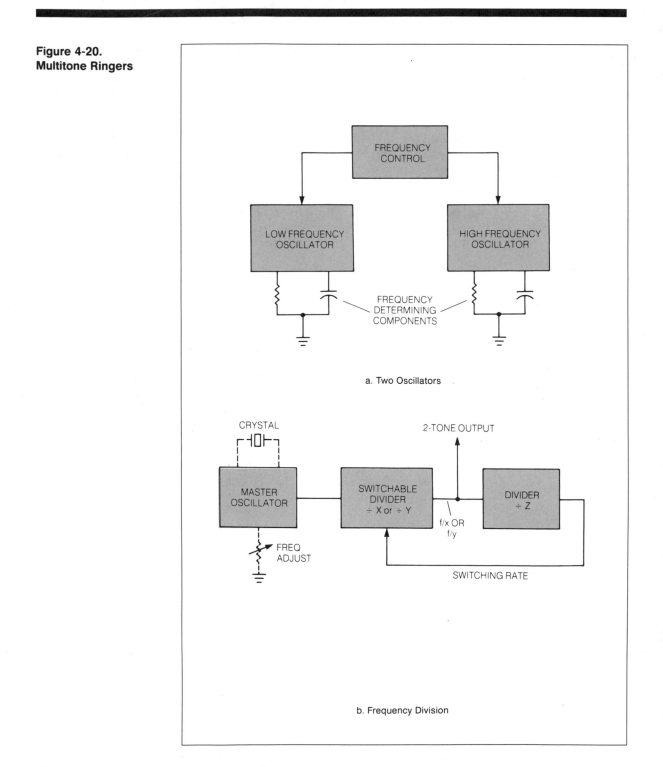

a. Two Oscillators

b. Frequency Division

Output Drivers

A good impedance match of the ringer output circuit is essential to produce maximum power output from the small input signal power available. A double-ended output stage will help provide an increased output over single-ended circuits.

The output driver (power amplifier) for the ringer circuit must match the transducer impedance to the circuit to produce the maximum audio output level with the small amount of input power available. A differential output driver can be used to increase the peak-to-peak voltage drive to the transducer. This technique is especially good for driving high impedance devices such as piezoelectric transducers. Instead of the more common single-ended output shown in *Figure 4-21a*, where the transducer is connected between one active output and ground, a double-ended output circuit has the load connected between two active outputs driven differentially in opposite phase (push-pull) as shown in *Figure 4-21b*. This method has the effect of doubling the peak-to-peak voltage applied to the transducer; thus, a much louder audio output is obtained.

**Figure 4-21.
Output Drivers**

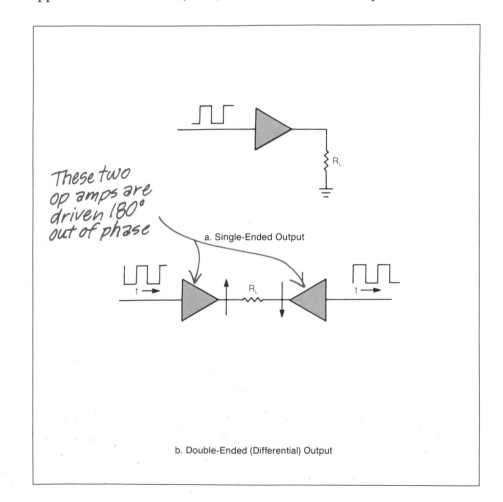

These two op amps are driven 180° out of phase

a. Single-Ended Output

b. Double-Ended (Differential) Output

ELECTRONIC RINGER IMPLEMENTATION EXAMPLE

Figure 4-22 shows an implementation of an IC multitone ringer that requires only two resistors, two capacitors, and the transducer external to the integrated circuit to perform the complete ringing function. All of the necessary circuitry including rectifiers and overvoltage protection is integrated into a single 8-pin dual-inline package. The operation of this circuit (a TCM1506 manufactured by Texas Instruments) is covered in the following discussion.

**Figure 4-22.
Texas Instruments
TCM1506 Ring
Detector and Ringer
Driver**

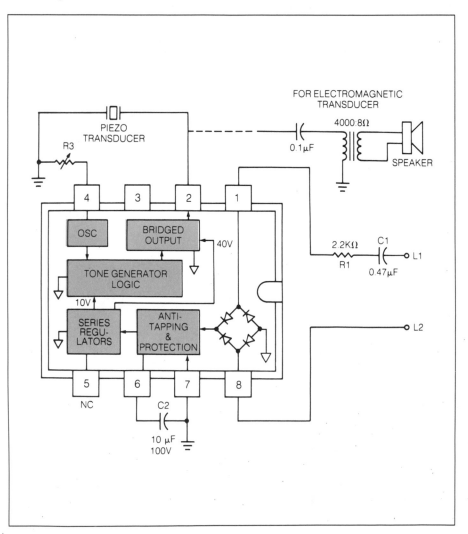

Rectification and Protection

The ac ringing voltage passes through C1 and R1 and is rectified by the on-chip diodes. Overvoltage protection for lightning spikes up to 1500 V for 200 microsecond duration is provided by a crowbar circuit. If a high-voltage pulse appears at the input to the chip, the crowbar circuit short circuits the input and dumps the excess power into resistor R1. The crowbar circuit then automatically resets. The rectified ac is filtered by an external capacitor, C2, and supplies the voltage regulator which powers the integrated circuit. A built-in oscillator supplies the timing frequency and can be adjusted with an external variable resistance, R3.

Tone Generation

The two-tone output signal is generated from a master oscillator by a programmable divider as was discussed for *Figure 4-20*. The shift rate between the high and low frequencies is controlled by another divider which counts cycles of the tone output to generate the low-frequency shift rate. Referring to *Figure 4-23*, with the master oscillator set to 48 kHz and the programmable divider set to divide-by-28, the tone output and input to the shift rate divider is 1,714 Hz. The shift rate divider (divide-by-128) counts 128 cycles at 1,714 Hz, then switches the programmable divider to divide by 32. Now the tone output and input to the shift rate divider is 1,500 Hz. The shift rate divider then counts 128 cycles at 1,500 Hz and returns the programmable divider to divide by 28. This cycle repeats as long as the ringing voltage is present on the line.

The shift rate is given by:

$$SR = \frac{1}{DSR(1/f1) + DSR(1/f2)}$$
$$= \frac{1}{\dfrac{128}{1714} + \dfrac{128}{1500}}$$
$$= 6.25 Hz$$

where,
    DSR is the shift divider rate ratio,
       f1 is the high output frequency,
       f2 is the low output frequency.

The frequency of the master oscillator in the circuit is controlled by adjusting R3 in *Figure 4-22*. Since the output frequencies are generated by division from a master oscillator, they retain their harmonic relationship of 28:32 or 1:1.14 regardless of the actual oscillator frequency.

**Figure 4-23.
TCM1506 Tone
Generator Block
Diagram**

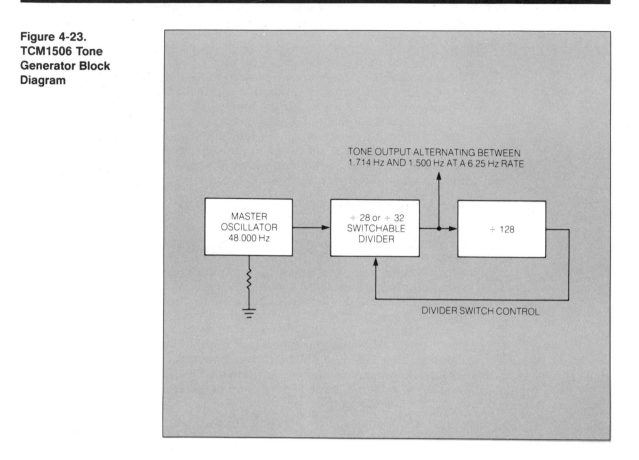

Output Driver

The output stage of the circuit in *Figure 4-22* is designed to drive a piezoelectric disc type transducer, or an electromagnetic transducer. The piezoelectric transducer is shown in solid lines. The output circuit is a single-ended design which enables the IC to generate 0 to 40 volts peak into an open circuit. If an electromagnetic transducer is to be driven, an extra 0.1-µF capacitor and a transformer which matches 4,000 ohms to an 8-ohm speaker is required. The connection is shown in dashed lines in *Figure 4-22*.

A COMPLETE INTEGRATED TELEPHONE

Integrated circuit dialing, ringing, and speech functions can be combined to form a complete, solid-state telephone which can provide such features as multitone ringing, tone or pulse dialing with memory and redial, and an active speech network free of bulky transformers or coils. *Figure 4-24* is the complete schematic for a solid-state telephone circuit.

**Figure 4-24.
A Complete Solid-
State Telephone.**
*(Courtesy Motorola,
Inc.)*

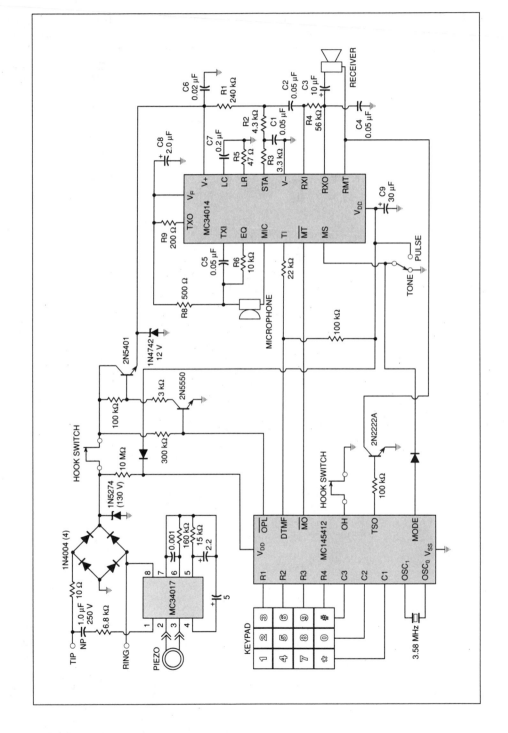

This circuit is completely line powered through a conventional polarity protection diode bridge using regular 1N4004 rectifier diodes. Transient protection is provided by a 1N5274 130-V zener diode. The MC145412 IC allows the circuit to function in either pulse or DTMF dialing modes. Tones are delivered directly to the MC34014 speech network discussed in Chapter 3. The speech network will amplify and couple DTMF signals to the local loop. When in pulse mode, a discrete transistor network consisting of a 2N5401 and 2N5550 will pulse the loop current. A 1N4742 12-V zener diode supplies secondary transient protection to the speech network during pulse dialing. An MC34017 ringer circuit connects to the line side of the telephone. The MC34017 contains its own diode rectifier, transient protection, and anti-tinkle circuitry. A piezoelectric sound element is driven by a push-pull type output driver.

WHAT HAVE WE LEARNED?

1. Pulse dialing, tone dialing, and ring detection and sounding can be accomplished with integrated circuits.
2. Integrated circuits need special protection from pulses of voltage called transients whose magnitude is so large that it exceeds the IC breakdown voltage and can damage the circuit.
3. Voltage transients are generated when current is interrupted in an inductive circuit.
4. Pulse dialers may provide electronic switching that is either in parallel or in series with the speech network.
5. The receiver is muted so the dialing pulses are not heard by the caller.
6. Muting may be accomplished by semiconductor switching devices or by relay contacts.
7. Typical tone dialing (DTMF) integrated circuits use a master oscillator and divide the frequency to obtain the tones for the row and column matrix.
8. Both single-tone and multitone ringer ICs are available.
9. Anti-tinkle circuits prevent the ringer from being energized during dialing.
10. Multitone ringers use a master oscillator and divide the frequency to obtain a low-frequency and a high-frequency tone for the multitone ringing signal.
11. Either electromagnetic or piezoelectric sound transducers may be used for ringing.

Quiz for Chapter 4

1. Memory is added to the telephone set to:
   a. allow more digits to be dialed
   b. allow faster dialing
   c. allow automatic redialing of the last number dialed
   d. none of the above

2. What is the function of the diode rectifier bridge in the line circuit?
   a. lower the voltage to the telephone electronics
   b. raise the voltage to the telephone electronics
   c. short out the line when the set is on-hook
   d. protect the set against polarity reversals on the line

3. What bad electrical effect happens when the switchhook is opened?
   a. a high voltage transient is generated
   b. the line is shorted out
   c. the line is opened
   d. no bad effects happen

4. Pulse dialing occurs at a rate of:
   a. 20 pulses per minute
   b. 10 pulses per minute
   c. 10 pulses per second
   d. 80 pulses per second

5. Pulse dialers must:
   a. mute the receiver in the set while dialing
   b. disconnect the transmitter while dialing
   c. short out the line while dialing
   d. all of the above

6. How many different tones may be produced by a four-column DTMF keypad?
   a. 2
   b. 8
   c. 4
   d. 16

7. What is an advantage of electronic ringers?
   a. louder volume
   b. smaller size
   c. good directionality
   d. greater weight

8. What functions are provided in a multitone ringing generator?
   a. anti-tinkle circuitry
   b. tone generation
   c. output amplifier
   d. all of the above

9. The DTMF generator in an electronic phone produces tones using a:
   a. LC circuit
   b. RC circuit
   c. digital divider circuit
   d. digital multiplier circuit

10. A piezoelectric transducer is:
   a. an Italian whistler
   b. a ceramic disk to produce sound
   c. a device used as a receiver
   d. a device used as a transmitter

Integrated Telephone Circuits

ABOUT THIS CHAPTER

Previous chapters have discussed the components found in conventional telephones, and how each component can be replaced with equivalent integrated circuits. But advances in electronics have not stopped there. Modern electronics now enables all standard telephone functions, as well as more sophisticated (or "feature") functions to be fabricated on a single IC chip. A whole new range of possibilities exist for telephones at little or no additional cost to the telephone user. This chapter will examine the features and applications of several advanced telephone integrated circuits.

THE MC34010 SINGLE-CHIP TELEPHONE

Stop for a moment and consider a totally integrated telephone circuit—a complete telephone with all necessary functions neatly packaged into a single IC. As a minimum, it would contain a dialer, ringer, speech network, and loop interface circuits. It would also require a few low-cost components to establish its operating parameters, yet satisfy all impedance and signalling requirements. Finally, it would automatically adjust for a wide range of loop lengths and operate properly on line voltages as low as 1.4 volts.

All telephone functions can be fabricated onto a single integrated circuit.

The Motorola MC34010 is an electronic telephone circuit which provides all of the functions listed in the preceding paragraph, as well as a microprocessor interface port for automated operations and features. This port not only allows for remote dialing commands from the microprocessor, but will also allow the microprocessor to interpret keypad input. Using this method, the microprocessor can communicate with computers or automated control systems right over the phone line. The elements of the electronic telephone are shown in *Figure 5-1*.

Line Interface

The line voltage regulator, or loop interface, is responsible for providing a constant level of voltage and current to the internal telephone circuitry. *Figure 5-2* is a diagram of the line interface circuit. Current is typically regulated through the pass transistor T1. It also selects the dc input resistance of the telephone. Capacitor C9 is a filter used to stabilize the regulated voltage.

Figure 5-1.
Block Diagram of a
Single-Chip Telephone
Circuit *(Courtesy*
Motorola, Inc.)

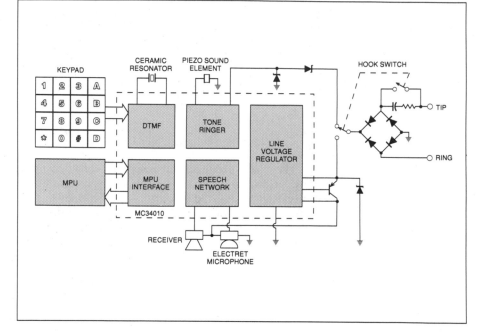

Figure 5-2.
Line Interface Network
Block Diagram
(Courtesy Motorola,
Inc.)

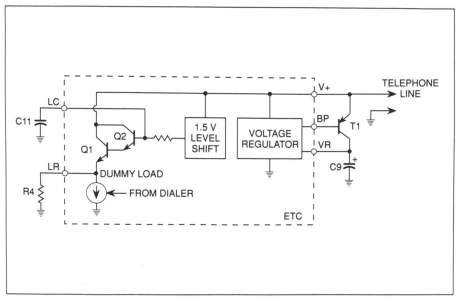

The line interface regulator controls the voltage and current delivered to the entire telephone circuit.

The telephone circuit is designed to operate properly with line voltages down to 1.4 volts. When the line voltage is below 3 volts, only the voltage regulator itself is operational to provide the necessary bias signals to the speech and dialer networks. When the line voltage exceeds 3 volts, the transistor network of Q1 and Q2 conducts and transfers excess line power to dc load resistor, R4. It is R4 which sets the dc input resistance of the circuit. Capacitor C11 acts as a dc load filter which prevents R4 from loading down any ac signals on the local loop.

Speech Network

The speech network interfaces and balances the 4-wire handset to the 2-wire telephone loop.

The speech network interfaces an electret transmitter and receiver to a two-wire telephone line. *Figure 5-3* shows the simplified block diagram of the speech network in the circuit.

Line current passed through transistor T1 is used to power an electret microphone. Resistor network R10, R11, R12, and R13 sets up proper bias between the microphone and the transmit amplifier. It is the fluctuation of current through this network that carries speech to the telephone line. A small portion of the transmit signal is delivered to a small amplifier which couples back into the receiver to provide sidetone.

Figure 5-3.
Speech Network Block Diagram (*Courtesy Motorola, Inc.*)

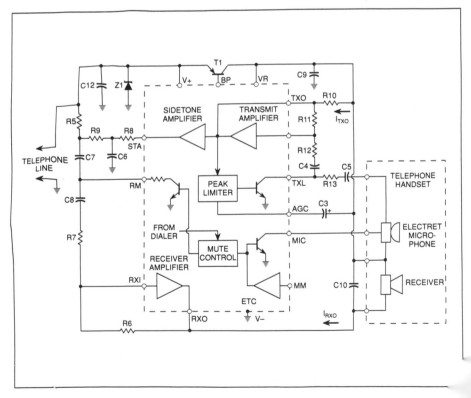

A peak detector and limiter is included to attenuate any loud transmit signal and restrict audio distortion to a low level. A mute signal from the internal dialer will shut down the microphone and receiver to suppress loud DTMF tones, as well as any annoying clicks due to hookswitch activity or keypad use (the MC34010 does not provide pulse dialing). The ac impedance of the speech network is essentially equal to the impedance of the receiver divided by the gain of the receiver amplifier.

Tone Dialer

A complete DTMF dialer network is included in the MC34010. It is fully compatible with 3×4 or 4×4 keypads. When a key is pressed, keypad comparator circuits define 3-bit row and column addresses of the selected key. Those 3-bit addresses are used by counter/encoders to select unique frequency division settings based on the master oscillator frequency. A varying 8-bit digital word is generated at the necessary frequency by the row and column encoders. Individual row and column D/A (digital to analog) converters translate 8-bit data words into their analog voltage levels. These synthesized tones are mixed together in an operational amplifier to produce the desired dual-tone output signal. *Figure 5-4* is a block diagram for the DTMF dialer.

Figure 5-4.
DTMF Dialer Block
Diagram *(Courtesy Motorola, Inc.)*

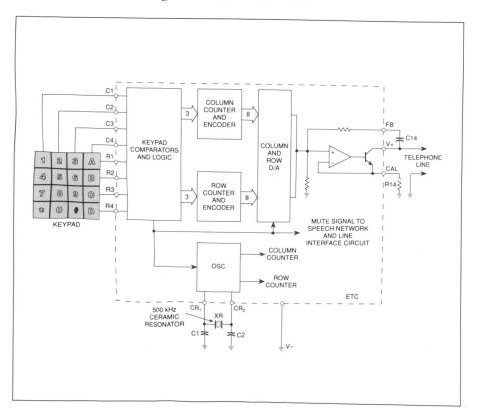

Tone dialing is the predominate dialing method in integrated telephone circuits.

The unique design used to generate row and column frequency signals can produce a tone which is accurate to within ±0.16%. As a result, an inexpensive 500-kHz ceramic resonator can be used instead of a crystal to supply the DTMF reference frequency. An oscillator with an accuracy of ±0.3% in this telephone circuit will support DTMF signals with an accuracy better than ±0.8%.

Ringer Network

The ringer network used in the Motorola MC34010 is very similar in approach to those individual ringer circuits presented in Chapter 4. *Figure 5-5* shows a block diagram for the Motorola MC34010 tone ringer. Ring signals occurring across the polarity protection bridge are full-wave rectified and clipped by zener diodes Z2 and Z3. When ring signal voltage exceeds the threshold voltage level set by R2, an 8/10 frequency divider will be enabled. This will supply an alternating two-tone, or warble, signal to a single-ended output buffer driving a piezoelectric sound element. When the ring voltage level drops below a lower ring threshold, the 8/10 frequency divider will be disabled and ring output will cease.

**Figure 5-5.
Tone Ringer Block
Diagram** *(Courtesy
Motorola, Inc.)*

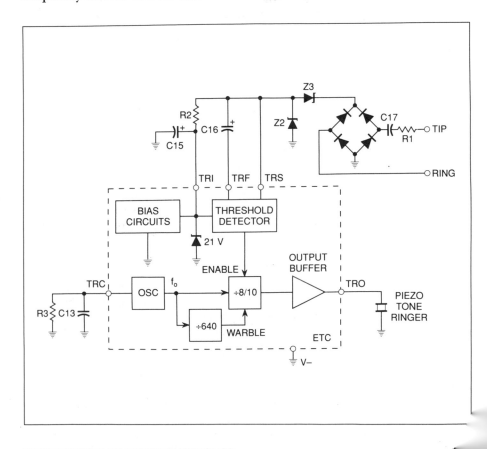

Ring threshold and ring oscillator frequency can be slightly adjusted to customize the ring tones.

Ring frequencies can be adjusted by external components. The base-frequency (f_o) of the free-running oscillator is determined by an RC network of R3 and C13. This network can be altered within very narrow limits to tailor the ring tones as desired.

USING A MICROPROCESSOR

Incorporating a microprocessor into a telephone circuit may seem rather excessive at first. After all, every function necessary to implement a telephone can be fabricated onto a single integrated circuit. The fact that a basic telephone set itself performs no logical or mathematical calculations can often add to this confusion.

Microprocessors allow many advanced features to be added to a telephone set with a minimum of additional cost.

In truth, the addition of a microprocessor does nothing to enhance the performance of an analog telephone circuit. The strength of a microprocessor lies in the broad range of additional functions that it can offer. Expanded number memory and recall, visual digit display (either through liquid crystal or light emitting diode displays), visual calender and clock, elapsed time and call-back indicators, automated redial, and answering system control are just some of the many features that microprocessors make possible. Telephones offering these enhanced features are often referred to as "intelligent telephones."

Choosing a Microprocessor

There are several key factors to remember when selecting a microprocessor for telephone applications. Perhaps the most important consideration is power. Since the microprocessor will have to be powered from the telephone line, it must be capable of operating at very low-voltage levels comparable to other telephone ICs. CMOS (Complementary Metal Oxide Semiconductor) and I²L (Integrated Injection Logic) devices will operate over a wide range of line voltages.

Another important consideration is microprocessor standardization. Although a custom-designed and manufactured microprocessor may prove more powerful than a standard model, the time, cost, and effort required to implement and program a custom device may be much greater than for an "off-the-shelf" device such as the Texas Instruments TMS7000 or the Motorola MC6800.

Microprocessor Interface

The Motorola MC34010 is equipped with an interface circuit specifically designed for connection to a microprocessor system. A block diagram for this interface is shown in *Figure 5-6*. It is incorporated into the DTMF dialer circuit and connected to the external microprocessor through six signal lines.

Figure 5-6. Microprocessor Interface Block Diagram *(Courtesy Motorola, Inc.)*

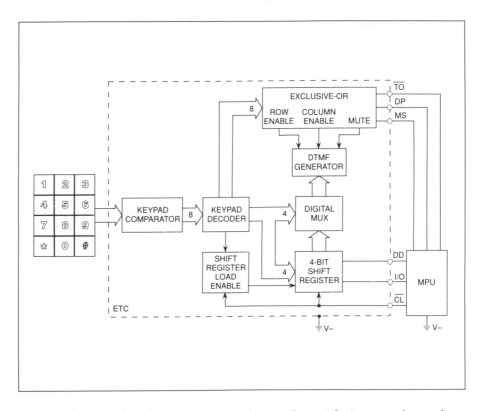

Each key on the dialpad is represented by a unique 4-bit binary number that can be read and interpreted by a microprocessor.

Each time a key is pressed (on either a 12- or 16-character keypad), row and column codes from the keypad comparators are translated into a 4-bit code through a keypad decoder. Every key has a unique code associated with it as shown in *Figure 5-7*. DTMF circuits use the key codes to load both programmable counters in order to generate appropriate tone combinations. The microprocessor interface circuit simultaneously loads the keypad code into a bidirectional 4-bit shift register for transfer to the microprocessor.

Transfer Method

Data is transferred serially to or from the microprocessor through an I/O pin from the shift register. The rate and timing of data transfer is controlled by clock pulses sent over the CL line by the microprocessor. It also manages the direction of data flow using the Data Direction (DD) pin. In this way, dialing activities of the telephone circuit can be supervised. Whether the microprocessor reads or writes key data will depend on what particular function it is performing and how it is programmed.

Figure 5-7.
Keypad Codes
(Courtesy Motorola,
Inc.)

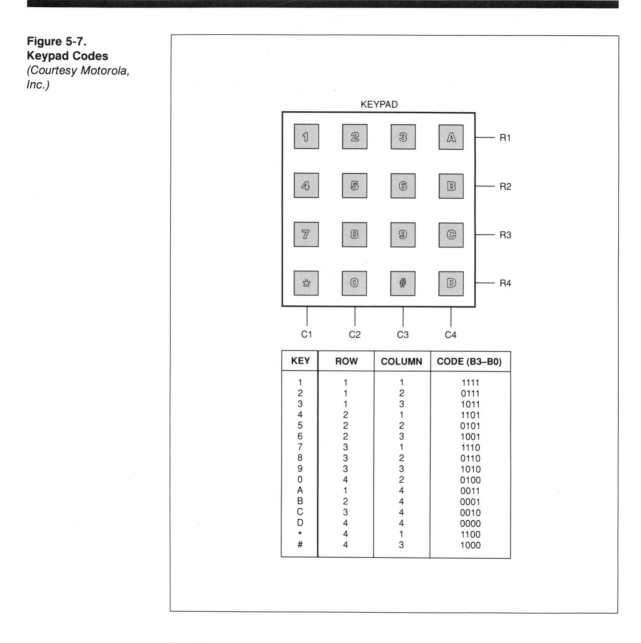

KEY	ROW	COLUMN	CODE (B3–B0)
1	1	1	1111
2	1	2	0111
3	1	3	1011
4	2	1	1101
5	2	2	0101
6	2	3	1001
7	3	1	1110
8	3	2	0110
9	3	3	1010
0	4	2	0100
A	1	4	0011
B	2	4	0001
C	3	4	0010
D	4	4	0000
*	4	1	1100
#	4	3	1000

Reading

If the microprocessor chooses to read information from the dialer for such things as programming telephone numbers or changing operating modes, the DD line is held logic low. When a key is pressed, its key code is channeled through to the DTMF generator. It is also loaded into the shift register. Every subsequent negative edge of the clock signal from CL will shift out one bit at a time through the I/O pin—most significant bit first. The speed at which the transfer takes place will depend on the clock rate at CL.

Writing

Any microprocessor system will require ROM (Read Only Memory) for permanent program storage, and RAM (Random Access Memory) for temporary data and variables.

If the microprocessor decides to write information to the dialer for purposes such as last number redial or automated control, DD will be held logic high. Four bits—most significant bit first—are loaded into the shift register through the I/O pin. One bit is loaded with each pulse on CL. In order to prevent anyone from entering erroneous data while the microprocessor is writing, it will hold the Tone Output, or TO line, in a logic high state. This signal disables tone generation until all 4 bits have been loaded. Once loaded, the microprocessor will return TO to a logic low and a tone will be produced.

Other Control Signals

The Motorola MC34010 generates two feedback signals that can be used by the microprocessor system. A dial pad signal (DP) becomes logic high when any valid key is pressed. When the key is released, the signal will return to a logic low level. Another status signal is the mute signal (MS). This signal becomes logic high when a tone is being produced and speech is muted. MS will return to a logic low after the tone is generated. Either of these signal lines may be used to signal the microprocessor or other control circuitry in the telephone.

APPLYING THE TELEPHONE CIRCUIT

The Motorola MC34010 electronic telephone circuit is designed to provide all of the features found in a more conventional telephone set with a minimum of external components. A typical application of the MC34010 is shown in *Figure 5-8*.

External resistors and capacitors establish operating characteristics of the telephone circuit such as input impedance, amplifier gain, and transient suppression.

With exception of the 300-ohm receiver element, there are absolutely no inductive components used in the electronic telephone. Impedance, gain, transient suppression, and filtering are all adjusted using off-the-shelf resistors and capacitors.

When the telephone is on-hook, switches S1 and S2 are positioned as shown in *Figure 5-8*. The telephone input impedance is now controlled by R1, C17, and Z3. Ideally, the telephone should have a very high impedance while on-hook. C17 will provide high impedance to dc and low frequencies in the voice passband. R1 will provide high impedance to all other frequencies. Zener diode Z3 shows a nonlinear impedance which helps to match the circuit as line voltage levels change. The polarity bridge (B1) is constructed using standard 1N4005 rectifier diodes.

Figure 5-8.
Electronic Telephone
Application Circuit
(Courtesy Motorola,
Inc.)

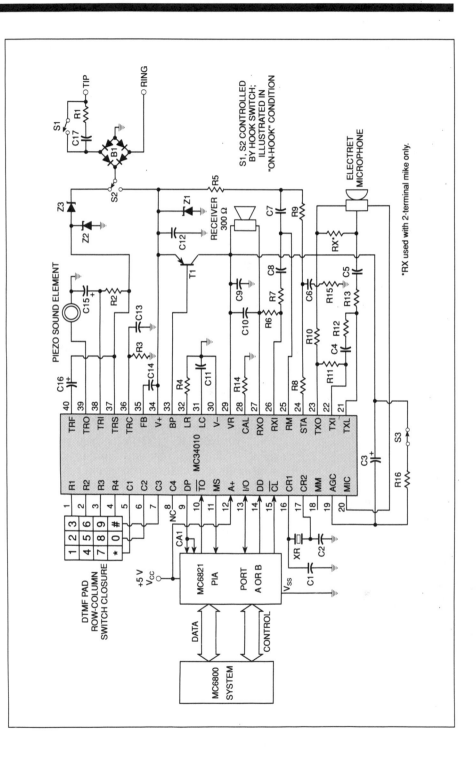

If the telephone is lifted off-hook, S1 shorts out R1 and C17 to reduce input impedance. Loop current will begin to flow from the central office. The dc resistance of the telephone circuit is now determined by R4 which conducts line current more than 10 milliamperes away from the circuit. The ac impedance while off-hook is controlled by the impedance of the receiver and receive amplifier gain:

$$Z_{offhook} = Z_{receiver}/A_{receiver}$$

Receiver gain is primarily controlled by R6. A larger value of R6 increases the receive gain, and vice versa.

Both ring start threshold and ring output frequencies can be adjusted through external components. R2 determines the ring start voltage level. An increase in R2 will decrease the voltage level required to start a ring cycle. The base frequency of the ring oscillator (f_o) is set with R3 and C13. It can be expressed as:

$$f_o = \frac{1}{(R3 \times C13) + (8 \times 10^6)}$$

where,
    f_o is the frequency in hertz.

The tone frequencies actually heard at the piezo sound element are fractions of the base frequency:

Upper frequency $= f_o/8$

Lower frequency $= f_o/10$

The warble rate—the rate at which upper and lower tones switch back and forth during ringing—is usually specified as:

Warble Rate $= f_o/640$

The ringer will operate properly with frequencies from 1 kHz to 10 kHz.

DTMF amplitude coupling to the local loop at V+ can easily be adjusted with R14. Lowering the value of R14 will increase the amplitude of the output tone. Since the relationship of all row and column tone amplitudes is fixed, a variation in R14 will alter all tone amplitudes.

The transmit output amplitude and transmit gain can be independently controlled in this telephone circuit. The transmitted voice signal reaching the local loop at V+ is restricted by R10. An increase in R10 will decrease the amplitude of the transmitted signal, and vice versa. Transmit gain is effected by R11, as well as the level of sidetone. By making R11 larger, a stronger signal will be driven through R10 to the local loop. The sidetone will tend to increase the transmit signal level even further.

Sidetone presence is controlled by the ratio of R9 and R5. R9 should always carry just a little bit more current than R5 to ensure that sidetone reaching the receiver will be in phase to the transmitter output. Resistors

R8, R15, and C6 form a phase shift network used to compensate for any phase shift introduced by the local loop. The sidetone signal is coupled into the receiver network through capacitors C7 and C8.

Transient protection is provided in several different ways. While on-hook, 30-volt zener diode Z2 protects the ringer circuit. When off-hook, switch S2 connects to the main portion of the circuit. Z1 is an 18-volt zener diode for speech network and dialer transient protection. Capacitor C3 will suppress any annoying clicks that might occur at the receiver. For a faster response, R16 may be connected across C3 by closing S3. Table 5-1 shows a detailed list of components used in the application of *Figure 5-8*, along with their typical values and descriptions.

Transient protection for integrated telephone circuits is usually provided by standard zener diodes.

**Table 5-1.
Motorola MC34010
External Components.**

Component	Nominal Value	Description
C1, C2	100 pF	Ceramic resonator oscillator capacitors.
C3	1 µF, 3.0 V	Transmit limiter low-pass filter capacitor: controls attack and decay time of transmit peak limiter.
C4, C5	0.1 µF	Transmit amplifier input capacitors: prevents dc current flow into TXL pin and attenuates low-frequency noise on microphone lead.
C6	0.05 µF	Sidetone network capacitor: provides phase-shift in sidetone path to match that caused by telephone line reactance.
C7, C8	0.05 µF	Receiver amplifier input capacitors: prevents dc current flow into RM terminal and attenuates low frequency noise in telephone line.
C9	2.2 µF, 3.0 V	VR regulator capacitor: frequency compensates the VR regulator to prevent oscillation.
C10	0.01 µF	Receiver amplifier output capacitor: frequency compensates the receiver amplifier to prevent oscillation.
C11	0.1 µF	Dc load filter capacitor: prevents the dc load circuit from attenuating ac signals on V+.
C12	0.01 µF	Telephone line bypass capacitor: terminates telephone line for high frequency signals and prevents oscillation in the VR regulator.
C13	620 pF	Tone ringer oscillator capacitor: determines clock frequency for tone and warble frequency synthesizers.
C14	0.1 µF	DTMF output feedback capacitor: ac couples feedback around the DTMF output amplifier which reduces output impedance.
C15	4.7 µF, 25 V	Tone ringer input capacitor: filters the rectified tone ringer input signal to smooth the supply potential for oscillator and output buffer.
C16	1.0 µF, 10 V	Tone ringer filter capacitor: integrates the voltage from current sense resistor R2 at the input of the threshold detector.

**Table 5-1.
(Cont.)***

Component	Nominal Value	Description
C17	1.0 µF, 250 V ac, Nonpolarized	Tone ringer line capacitor: ac couples the tone ringer to the telephone line; partially controls the on-hook input impedance of the telephone.
R1	6.8 kΩ	Tone ringer input resistor: limits current into the telephone line and partially controls the on-hook impedance of the telephone.
R2	1.8 kΩ	Tone ringer current sense resistor: produces a voltage at the input of the threshold detector in proportion to the tone ringer input current.
R3	200 kΩ	Tone ringer oscillator resistor: determines the clock frequency for tone and warble frequency synthesizers.
R4	82 Ω, 1.0 W	Dc load resistor: conducts all dc line current in excess of the current required for speech or dialing circuits; controls the off-hook dc resistance of the telephone.
R5, R7	150 kΩ, 56 kΩ	Receiver amplifier input resistors: couples ac input signals from the telephone line to the receiver amplifier; signal in R5 subtracts from that in R9 to reduce sidetone in receiver.
R6	200 kΩ	Receiver amplifier feedback resistor: controls the gain of the receiver amplifier.
R8, R9	1.5 kΩ, 30 kΩ	Sidetone network resistors: drives receiver amplifier input with the inverted output signal from the transmitter; the phase of signal in R9 should be opposite that in R5.
R10	270 Ω	Transmit amplifier load resistor: converts output voltage of transmit amplifier into a current that drives the telephone line; controls the maximum transmit level.
R11	200 kΩ	Transmit amplifier feedback resistor: controls the gain of the transmit amplifier.
R12, R13	4.7 kΩ, 4.7 kΩ	Transmit amplifier input resistors: couples signal from microphone to transmit amplifier; controls the dynamic range of the transmit peak limiter.
R14	36 Ω	DTMF calibration resistor: controls the output amplitude of the DTMF dialer.
R15	2 kΩ	Sidetone network resistor (optional): reduces phase shift in sidetone network at high frequencies.
R16	100 Ω	Hook switch click suppression current limit resistor (optional): limits current when S3 discharges C3 after switching to the on-hook condition.
R_x	3 kΩ	Microphone bias resistor: sources current output from VR to power a 2-terminal electret microphone; R_x is not used with 3-terminal microphones.

* *Courtesy Motorola, Inc.*

The Motorola MC34010 is just one of a variety of specialized telephone integrated circuits developed for the telecommunications industry. While this device offers convenience and simplicity over more conventional telephone assemblies, another type of telephone is growing in popularity—speakerphones.

SPEAKERPHONES

Speakerphones contain a transmitter and receiver (almost always a speaker—thus the name ''speakerphone'') built into the main housing of the telephone set as shown in *Figure 5-9*. This eliminates the need for a handset, although a handset is usually added for privacy and convenience. By mounting a transmit and receive element in exposed locations on the telephone, it is possible to speak or listen to a caller from just about anywhere within a room. This also allows groups of people to participate in the same conversation without requiring several extension telephones off-hook on the same loop. Multiple extensions would load down signals and make the conversation more difficult for everyone to follow.

**Figure 5-9.
View of a Basic
Speakerphone Set**

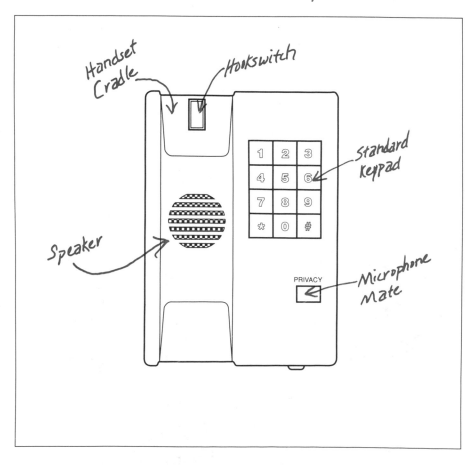

Speakerphones, also known as "comfort phones" or "hands-free" phones, are certainly not new devices. They have existed in one form or another for decades, although several drawbacks have kept many of them out of the commercial market until just recently.

Once bulky, expensive, and unreliable, speakerphones are making extensive use of specialized ICs to improve their performance.

The principle problem with older speakerphones is feedback. Speaker signals travel freely through surrounding air, and can be picked up again by the microphone as transmit signals. These "transmitted" signals would be amplified and delivered to the local loop. A portion of this would be sent back to the speaker as sidetone, only to be coupled to the microphone again. Feedback is an important component of most oscillator circuits. In this case, however, the self oscillation produced by voice feedback causes a loud, incredibly annoying wail or whine. As a result, full-duplex operation of a speakerphone is impossible. Calling and called parties must take turns speaking and listening in half-duplex mode. Complex sensing and switching circuitry had to be added to switch between transmit and receive modes. The resulting increase in bulk and cost often placed speakerphones beyond the range of the commercial market. A new generation of speakerphones uses advanced integrated circuits to overcome traditional disadvantages.

Speakerphone Functions

In order for a speakerphone to operate properly, there are several important functions that must be performed. *Figure 5-10* is a simplified block diagram of a speakerphone speech network.

Detection and switching circuitry are important functions in the speakerphone speech network.

The speakerphone must be capable of amplifying the transmit and receive signals delivered from the local loop through the balance network. Since a caller may be fairly far away from the phone, transmit and receive amplifiers provide substantial gain. Their signal levels are constantly monitored and checked against levels of background noise. Monitors control the actions of a switching circuit. The switching circuit will turn transmit or receive attenuators on or off depending upon which party is speaking at any particular moment. Balance network circuitry performs a 4-wire to 2-wire conversion of the speech signals. Notice the differences between this speech network, and the telephone speech network discussed in Chapter 3.

INTEGRATED SPEAKERPHONE CIRCUITS

All of the circuits required to perform amplification, level detection, attenuation, switching, and hybrid functions can now be fabricated on a single integrated circuit. This level of integration offers a smaller, more reliable circuit which is easier to assemble and test, as well as a tremendous cost savings. Motorola and Exar both manufacture speakerphone ICs.

Figure 5-10.
Block Diagram of a
Speakerphone Speech
Network

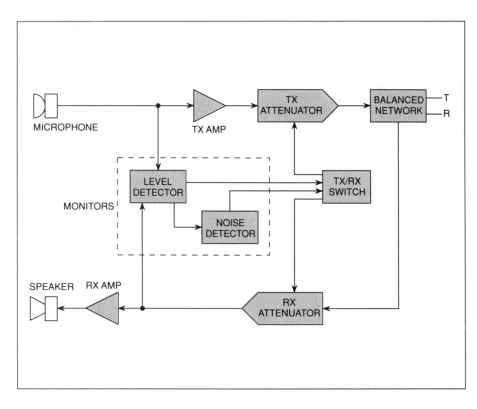

A typical speakerphone IC contains amplifiers, attenuators, level detectors, switching, and control circuitry.

Figure 5-11 is a block diagram for the Motorola MC34118. It contains all of the circuits needed to implement a complete speakerphone speech network. It supplies a transmit amplifier, level and background noise detectors, transmit and receive attenuators, and switching/control circuitry.

Attenuators

There are two attenuators in the speech path—one for transmit and one for receive. They operate in a complementary fashion under the direction of the attenuator control circuit, so one attenuator is always off while the other is on. This reversing action is what makes half-duplex operation possible. Although the attenuators do not create an open circuit when activated, they will provide as much as –46 dB of attenuation to the signal.

The attenuator control circuit interprets four level inputs; two from background noise monitors, and two from level detectors as detailed in *Figure 5-12*. Switching characteristics are also effected by a volume control setting, a dial tone detector, and an automatic gain circuit. Each of these factors let the attenuator control circuit decide when to transmit and receive.

**Figure 5-11.
Motorola MC34118
Block Diagram**
*(Courtesy Motorola,
Inc.)*

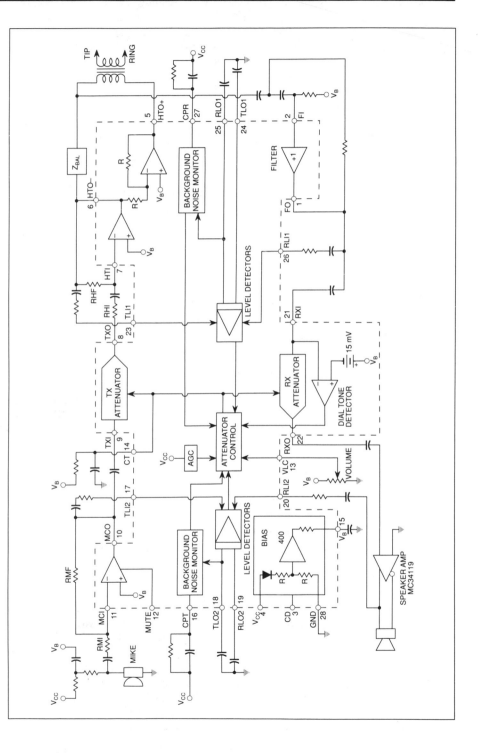

**Figure 5-12.
Control Circuit
Diagram** *(Courtesy
Motorola, Inc.)*

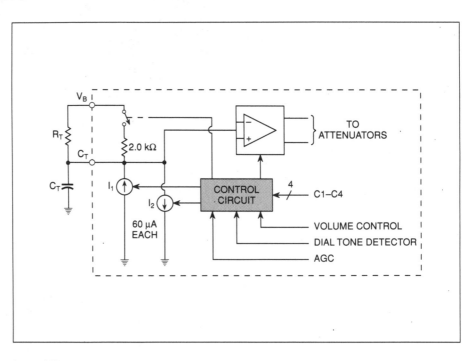

Level Detectors

There are two sets of two level detectors in the MC34118; two in the receive path and two in the transmit path. Level detectors are little more than high-gain comparators. The output states will change when an input is greater or less than a reference signal. The switching sensitivity of each level detector can be set independently using individual RC networks. The attenuator control circuit interprets each of these inputs to determine the proper operating mode for the speakerphone. A more detailed diagram of a level detector set is shown in *Figure 5-13.*

Background noise monitors, which are designed to distinguish the characteristics of speech from background noise, are also added. Since speech is a fluctuating signal and background noise is typically constant, under most circumstances it is a simple matter to tell two conditions apart.

Amplifiers

Speech signals picked up by an electret microphone will be boosted as much as 80 dB by a single-stage amplifier. This ensures that even the weakest detected signal will be transmitted. A mute option is added to provide privacy for the caller on demand. If the caller wishes to discuss something aloud without having that speech overheard by the called party, a mute key can be pressed. A logic high condition at the mute pin will disable the microphone amplifier and attenuate any transmitted signals. A logic low condition will restore normal operation. *Figure 5-14* shows a detailed diagram of the microphone amplifier and mute circuit.

Fig 5-13.
Level Detector Set
(Courtesy Motorola,
Inc.)

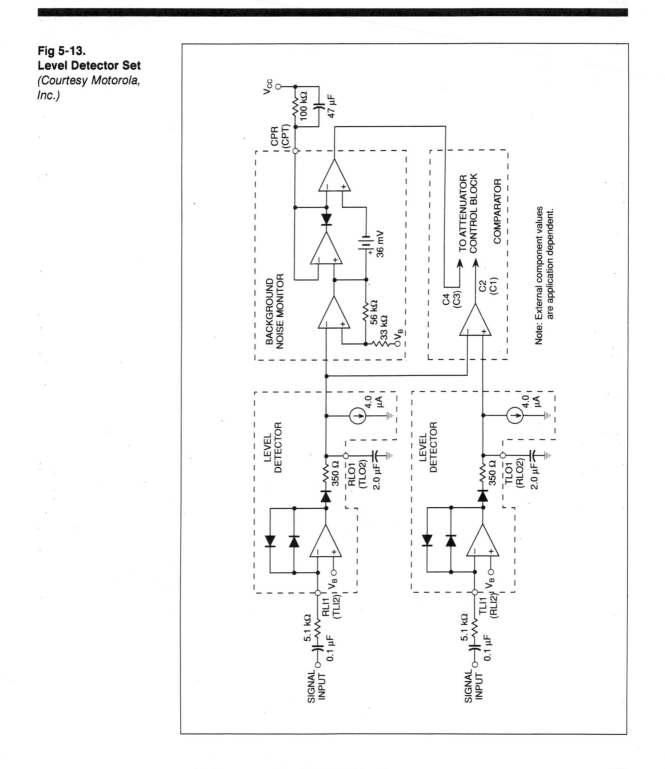

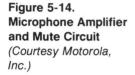

Figure 5-14.
Microphone Amplifier
and Mute Circuit
(Courtesy Motorola,
Inc.)

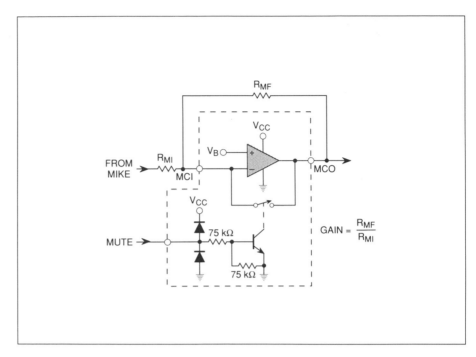

Amplifiers are used to construct the MC34118 hybrid network which provides the four-wire to two-wire local loop interface. Gain in the hybrid can be set with external components. Finally, an operational amplifier is used as a filter in the receive circuit. By properly selecting an appropriate configuration of resistors and capacitors, a bandpass filter can be formed. Such a filter will attenuate low-frequency noise such as ac line noise, as well as high frequency signals which may cause feedback problems.

Applying the MC34118

A basic hands-free speakerphone is shown in *Figure 5-15*. The Motorola MC34118 forms the heart of the network. Operating parameters are set with a minimum of external resistors and capacitors. This particular circuit uses a 25-Ω 300-milliwatt speaker. Since the MC34118 can not supply enough power to drive the speaker directly, an MC34119 audio amplifier is included.

Even though the circuit of *Figure 5-15* is a functional application of the MC34118, it suffers from several important drawbacks. First, there is no ringer. This means that there is no way of signalling an incoming call. Second, there is no means of dialing, so it is impossible to signal another telephone. Finally, no handset is available. Users are limited to use of the open speaker and microphone only. All of these disadvantages are eliminated with the complete speakerphone/telephone application of *Figure 5-16*.

**Figure 5-15.
Motorola MC34118
Application Circuit**
*(Courtesy Motorola,
Inc.)*

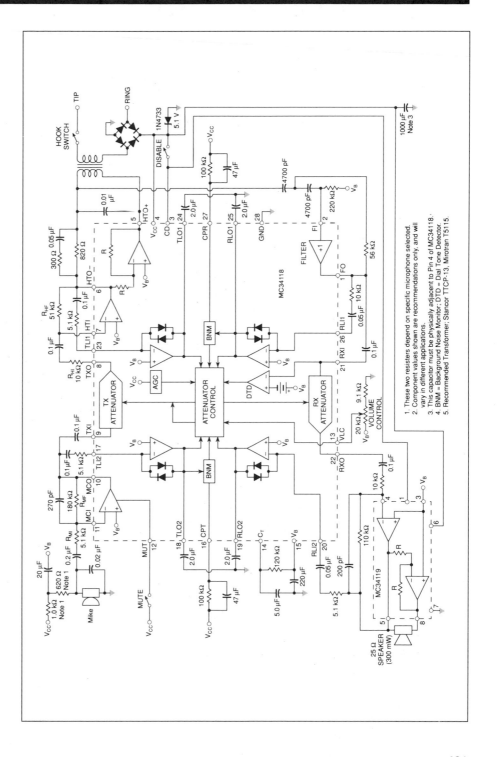

**Figure 5-16.
Pulse/Tone Comfort-
Phone with Memory**
*(Courtesy Motorola,
Inc.)*

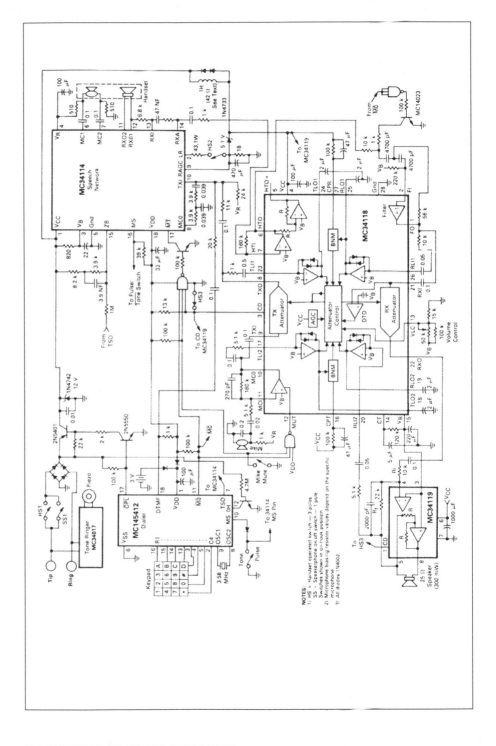

The line-powered circuit of the MC34118 and MC34119 are essentially the same as *Figure 5-15*, but a variety of major improvements have been made. An MC34017 ringer circuit will provide a warble ringing signal to its accompanying piezo sound element. Dialing is available in either pulse or tone mode through use of an MC145412 dialer. Notice a 3-volt battery next to the dialer. This provides battery backup for numbers stored in memory. A diode is added in line with the battery to allow current flow from the battery only. It prevents forced battery charging whenever line voltage exceeds battery voltage. An MC34114 serves as a secondary speech circuit to support a handset. In this way, the telephone can switch from handset to hands-free operation and back again on demand.

Adding Intelligence to the Speakerphone

Although the speakerphone circuit is not equipped with a microprocessor interface, it would be possible to add a microprocessor as shown in *Figure 5-17*. *Figure 5-17* illustrates a block diagram of the Pulse/Tone Speakerphone shown by *Figure 5-16*. Circuitry for the microprocessor, its program and workspace memory, and an alphanumeric visual display have been added. Notice that the keypad is now used to enter row and column signals to the microprocessor. It senses on-hook or off-hook conditions, and decides whether the key entries should be sent on to the dialer, or interpreted as a command code to update the visual display or other operating parameters such as date or time.

**Figure 5-17.
Intelligent
Speakerphone Block
Diagram**

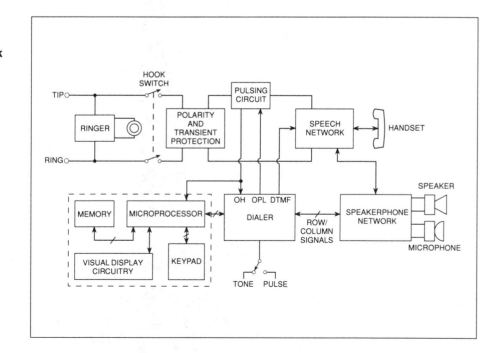

WHAT HAVE WE LEARNED

1. All functions necessary to implement a telephone can be fabricated onto a single integrated circuit.
2. Adding a microprocessor to a telephone circuit allows many intelligent features to be included for little or no additional cost.
3. Integrated telephone circuits may be able to operate properly on line voltages as low as 1.4 volts.
4. CMOS and I²L devices are often used in integrated telephone circuits because of their wide operating voltage range and noise immunity.
5. Each of the functions in telephone ICs can be optimized with a minimum of resistors and capacitors.
6. Any microprocessor circuit will need some minimum amount of permanent (ROM) and temporary (RAM) memory.
7. Speakerphones, or hands-free phones, allow the caller to speak or listen from just about anywhere in the room.
8. Feedback has always been a problem in speakerphone designs.
9. The switching, amplification, and detection circuitry necessary for a speakerphone can be fabricated onto a single integrated circuit.
10. A microprocessor can be added to just about any integrated telephone circuit.

Quiz for Chapter 5

1. Which function is not required in an integrated telephone circuit ?
   a. Regulator
   b. Dialer
   c. Visual Display
   d. Speech Network

2. What does a microprocessor require in order to work with a telephone circuit ?
   a. ROM
   b. RAM
   c. A specialized interface circuit in the telephone
   d. a and b above

3. The regulator in an integrated telephone circuit controls:
   a. amplitude.
   b. current.
   c. voltage.
   d. b and c above.

4. CMOS stands for:
   a. Covered Metal On Silicon.
   b. Conventional Metal Oxide Silicon.
   c. Complementary Metal Oxide Semiconductor.
   d. none of the above.

5. I²L stands for:
   A. Internal Integrator Logic.
   b. Iterative Injection Logic.
   c. Integrated Injection Logic.
   d. none of the above.

6. Transient protection for integrated telephone circuits is typically provided by:
   a. bridge rectifiers.
   b. zener diodes.
   c. inductors.
   d. varistors.

7. Integrated telephone and speakerphone circuit performance is optimized with:
   a. resistors.
   b. inductors.
   c. tunnel diodes.
   d. capacitors.
   e. a and d above.

8. Speakerphones operate in:
   a. full-duplex mode.
   b. half-duplex mode.
   c. open-duplex mode.
   d. computer mode.

9. Microprocessors cannot be interfaced to speakerphones:
   a. True
   b. False

10. Integrated telephones provide more gain to transmitted and received signals than speakerphones:
    a. True
    b. False

Digital Transmission Techniques

ABOUT THIS CHAPTER

This chapter explains the principles of using digital techniques to transmit telephone signals. It describes Pulse Code Modulation (PCM) and Time Division Multiplexing (TDM) systems, and compares them with analog systems. Digital transmission techniques deal with information all in digital form. To better understand the techniques used, let's cover some key concepts about digital signals.

DIGITAL SIGNALS

The digital signal consists of a sequence of bits which have two values.

A digital signal, as we briefly described in Chapter 1, is made up of a combination of separate parts called bits. A bit can have only two values, 0 and 1. The 0 and 1 values may represent many different two-valued conditions; on-off, full-current–no-current, high-voltage–low-voltage, true–false, etc. To understand this further, look at *Figure 6-1a.*

Single-Bit Digital Signals

A digital signal in a circuit may be generated by mechanically closing and opening a switch at controlled time intervals. The voltage output when plotting against time yields a sequence of digit signals.

The digital signal shown in *Figure 6-1a* is changing from the 0 level to the 1 level and back again as time passes. The 0 and 1 state could also be represented by voltages as shown in *Figure 6-1b*, +5 volts for the 1 level and 0 volts for the 0 level. Then the waveform of *Figure 6-1b* could be generated by measuring the voltage at switch S_o in the circuit of *Figure 6-1c* and changing the switch position from open to closed at particular intervals of time. To generate the waveform shown, S_o is held open for 1 second, closed for 1 second, open for 4 seconds, closed for 2 seconds, open for 3 seconds, closed for 1 second, open for 1 second, closed for 1 second, and open for 2 seconds.

Although this example is useful for understanding digital signals, the reader should be aware that moving switch S_o of *Figure 6-1c* back and forth to generate the signals and measuring the digit states with meters is much too slow for real-world digital systems. Electronic circuits recognize the levels and sense and transfer the digital states between circuits at a few billionths of a second (1×10^{-9} seconds) so that millions of transactions occur in the wink of an eye.

The line L1 in *Figure 6-1c* could be a control line in any digital system, or being more specific, it could be a control line in a telephone switching network. By continually sensing the signal level on L1, the state of the control line can be determined.

**Figure 6-1.
Single-Bit and
Multiple-Bit Serial
Digital Signals**

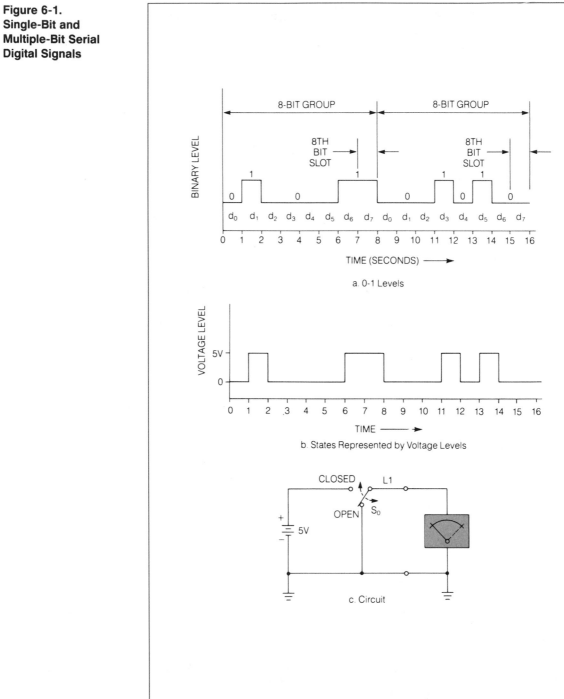

a. 0-1 Levels

b. States Represented by Voltage Levels

c. Circuit

Multiple-Bit Digital Signals—Serial

Serial bit transmissions may be thought of as a continuous stream of highs and lows (voltage and no-voltage) passing at a fixed rate of speed down an electrical line.

If the information contained on L1 in *Figure 6-1c* needs to be moved or transmitted to another location, it would be transmitted in serial form. Serial transmission can be visualized by letting the waveform of digital information pass by a fixed position with time. For example, if in *Figure 6-1a* the fixed position were the Y (level) axis, and the waveform were moved to the left as time passes; for the first second a 0 level would be passing the Y axis, then the level would change to a 1 and remain there for one second. Next the level would change to a 0 and remain at 0 for four seconds, etc. Therefore, if an observer were watching the bits go by the Y axis, the following code would be seen in the first eight seconds:

0 1 0 0 0 0 1 1

In the next eight seconds, the code would be:

0 0 0 1 0 1 0 0

Multiple-Bit Digital Signals—Parallel

Parallel bit transmission may be thought of as two or more continuous streams of highs and lows, passing at a fixed rate of speed down two or more lines simultaneously. In three-bit processing three levels exist.

Suppose now that instead of one switch as in *Figure 6-1c,* there are three switches as in *Figure 6-2a*. Each switch is switched every second as before, but the pattern of 1s and 0s that they generate is independent. However, all switching is started at the same time so every second, if a switch changes, it changes at the same time as the other switches.

The 0 and 1 patterns generated by the three switches in a particular case are shown in *Figure 6-2b*. As shown in *Figure 6-2a*, these patterns were detected with a meter on each line measuring the voltage the same way as for *Figure 6-1c*. They will be used to describe another digital technique—parallel bit processing.

The patterns in *Figure 6-2b* are going to be visualized in serial form as in *Figure 6-1a*. The fixed position again is the Y axis. However, now all three waveforms are going to be moved past the Y axis at the same time. Each pattern is going to be assigned a bit position: d_0 for the S_0 pattern, d_1 for the S_1 pattern, and d_2 for the S_2 pattern.

At the start (time = 0 seconds), $d_0 = 0$, $d_1 = 0$ and $d_2 = 0$. The code at time zero is 000. One second later the code changes to $d_0 = 1$, $d_1 = 0$, and $d_2 = 0$. All of the bits of the code are moving past the Y axis at the same time so the bits are appearing (being output) in parallel. This is parallel bit processing in digital systems.

The three bits can exist in eight different combinations. Each combination or code may represent a specific number or character.

The binary signals (codes) generated for the different combinations of d_0, d_1, and d_2 at the different time periods are listed in *Figure 6-3*. Column 2 shows that eight different codes can be generated using all combinations of three bits. In general, if there are N bits, there are 2^N different codes (bit patterns of 1s and 0s) that can be used to identify 2^N different conditions. As shown in *Figure 6-3*, the eight codes could represent numbers from 0 to 7, or they could represent a total of 8

Figure 6-2.
Parallel Digital Signals

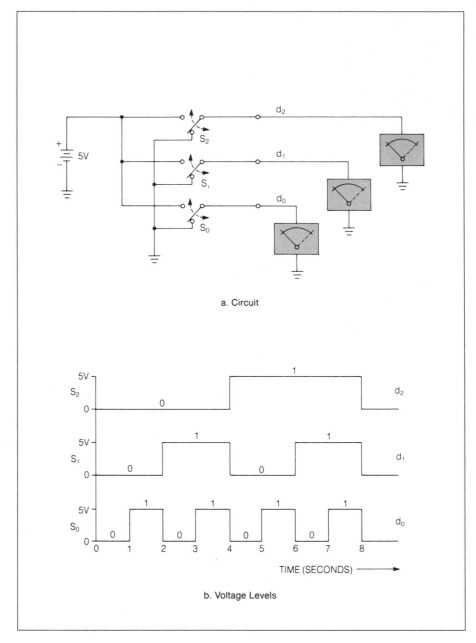

a. Circuit

b. Voltage Levels

different commands and/or characters. Any number of different conditions can be represented by just adding more positions to the bit pattern. An 8-bit group happens to be a convenient one used often in telephone systems. With it, 256 different conditions can be represented.

**Figure 6-3.
Binary Codes
Representing Different
Conditions**

After Time (Secs)	Binary Signals			Decimal Number Equivalent	Other Possible Equivalences
	d_2	d_1	d_0		
0	0	0	0	0	A
1	0	0	1	1	STOP
2	0	1	0	2	GO
3	0	1	1	3	B
4	1	0	0	4	OFF
5	1	0	1	5	ON
6	1	1	0	6	RIGHT
7	1	1	1	7	LEFT
EIGHT DIFFERENT TIMES	EIGHT DIFFERENT SIGNAL PATTERNS			EIGHT DIFFERENT DECIMAL NUMBERS	EIGHT DIFFERENT CHARACTERS OR COMMANDS

How are Numbers Represented?

In the decimal system 10 digits are used, 0 through 9. It is called a base 10 system. In the decimal system, numbers are grouped right to left with the first digit representing the ones place (10^0), the second digit the tens place (10^1), and so on. Each place increases in value by a power of 10.

The binary system is a base 2 system. In most cases, numbers are grouped right to left the same as the decimal system. The right-most bit, d_0, is in the ones place (2^0), the second bit, d_1, is in the twos place (2^1), the third bit, d_2, is in the fours place (2^2), the fourth bit is in the eights place (2^3), and so on. Each place increases in value by a power of 2. The two systems are summarized in *Table 6-1*.

In the decimal numbering system, each position going from right to left increases its value by a power of 10. In the binary numbering system, each position increases in value by a power of 2.

**Table 6-1.
Comparison of Place
Values**

	DECIMAL (Base 10)				BINARY (Base 2)				
PLACE	4	3	2	1	5	4	3	2	1
VALUE	1000	100	10	1	16	8	4	2	1

By using these place values wherever a 1 appears in the code of *Figure 6-3* and adding them together, the equivalent decimal number can be obtained. The evaluation of the code 011 is as follows:

$$(1 \times d_0) + (1 \times d_1) + (0 \times d_2) =$$
$$(1 \times 1) + (1 \times 2) + (0 \times 4) =$$
$$1 + 2 + 0 = 3$$

Representing Information with Binary Codes

To represent information, the bits are grouped together in several different ways. One way uses each 8-bit group to represent a number or letter. This eight bit group is called a byte.

Another way assigns a specific bit to indicate a particular function or state.

When binary codes represent information, whether they are formed serially or in parallel, the information may be carried in a number of ways. The first way is to use the complete group of bits.

Figure 6-1a illustrated an 8-bit group which is a common grouping for telephone systems as well as digital systems. All of the bits in the group are needed to identify the information. The 8-bit group forms a code to identify a unique condition of the information that has been digitized. The bits in each 8-bit group are numbered from d_0 to d_7 as shown in *Figure 6-1a*; d_0 is called the least significant bit (LSB) and d_7 the most significant bit (MSB).

In addition to using the complete group of bits as a code, a second way is to use specific individual bits within a group to control specific functions or to indicate a particular status. A typical example in digital telephone systems is to assign the state of one bit in a group to indicate whether a telephone is on-hook or off-hook. An example using *Figure 6-1a* again will illustrate this technique.

Bit d_7 of an 8-bit group is used to indicate the status of the subscriber telephone set. When the handset is off-hook, d_7 equals 1; when it is on-hook, d_7 equals 0. Therefore, in *Figure 6-1a* when the first 8-bit group of bits comes by in serial form, d_7 equals 1 and the system detects that the handset is off-hook. When the second 8-bit group comes by, d_7 equals 0 and the on-hook condition is identified. Other bit positions can be used similarly to identify two conditions of a control signal.

A third method subdivides the group into subgroups, each subgroup representing a character in a group of characters.

A third way to carry information is to use subgroups of bits. Suppose the 8-bit group code is divided into subgroups as follows:

```
010    1100   1
 A       B    C
```

In this technique, the first three bits of the code, subgroup A, could be used to identify one of eight different regional centers. Subgroup B, the next four bits, could be used to identify something entirely different—one of 16 different trunk lines, for example. The last bit, subgroup C, still could be used to indicate that a handset is on-hook or off-hook.

SIGNAL CONVERSIONS

Because the early telephone systems were completely analog and a large portion of the system today is still analog, there is a need for conversion from analog signals to digital signals so they may be transmitted using digital techniques. When the signal arrives at the destination it must be converted back to its analog form. Let's briefly review both of these concepts.

Analog to Digital

To convert an analog signal to digital, the range of analog voltage levels is divided into discrete levels, each represented by a unique 8-bit code.

Figure 6-4a shows the basic principle of an analog to digital (A/D) converter. The input signal is a continuously varying analog of the speech input to the telephone transmitter. At predetermined times, the input analog signal is sampled and the voltage value is converted to an equivalent digital code. In this example, an 8-bit code is used. The code comes out of the converter in parallel each time the input is sampled and its value represents the sampled voltage value. The parallel codes are converted to serial form for transmission. After arriving at the destination, they are reconverted to parallel form to feed a digital to analog (D/A) converter.

**Figure 6-4.
Signal Conversions**

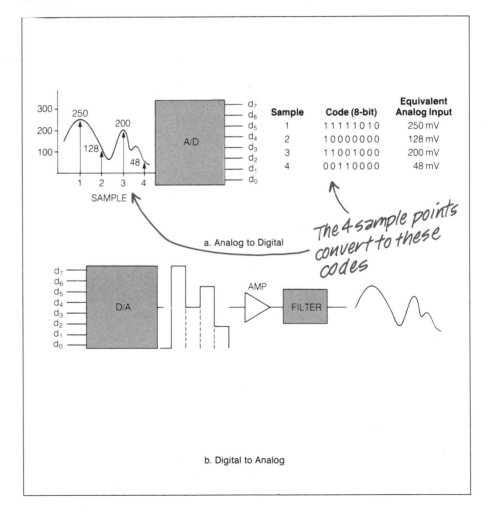

Sample	Code (8-bit)	Equivalent Analog Input
1	1 1 1 1 1 0 1 0	250 mV
2	1 0 0 0 0 0 0 0	128 mV
3	1 1 0 0 1 0 0 0	200 mV
4	0 0 1 1 0 0 0 0	48 mV

a. Analog to Digital

The 4 sample points convert to these codes

b. Digital to Analog

Digital to Analog

From the parallel 8-bit code input, the D/A converter outputs a voltage level for each code as shown in *Figure 6-4b*. This voltage level remains constant for the sample period and the output waveform has stepped levels. Passing the step-level output through an amplifier and filter restores the signal very nearly to its original smooth and continuously varying shape. The more samples taken in a given period of time by the A/D converter, the more accurate the reproduction of the signal at the output of the D/A converter.

To convert a digital signal to analog, the D/A (analog to digital) converter takes the digital signal and outputs an approximately similar, but stepped signal to an amplifier/filter network, which smooths the output and provides a nearly perfect replica of the signal.

ADVANTAGES AND DISADVANTAGES OF USING DIGITAL SIGNALS

Changing the analog telephone network to digital has progressed rapidly since 1962, when the first digital transmission system was installed by the Bell System in Chicago. This rush to convert the analog network to digital technology has not arisen because of an overwhelming demand for transmission of data between machines. Instead, the advantages of digital transmission techniques applied to speech signals have proved to be so numerous as to make the continued installation of analog facilities a poor business choice. What are these advantages? Let's look at some of them.

Advantages

Lower Cost Using Common Circuitry

The telephone transmission network and switching equipment for digital systems can use the same types of integrated circuit logic used in digital computers. This circuitry has been declining in cost by a factor of two every three years for about the last twenty years. The cost of analog circuits has not declined as fast. As a result, digital systems cost less and, as more systems are installed to increase the volume, the cost will continue to go down. There are currently no more analog switching machines being designed, except possibly very small ones. The net result is a reduced volume which has increased the cost of analog equipment.

For telephone applications, digital electronic techniques are becoming increasingly more cost-effective than analog.

Common Circuit Functions

When both the transmission technique and the switching system are digital, it is possible to integrate transmission and switching such that many of the traditionally required interface circuits, such as the two-wire to four-wire conversion shown in *Figure 6-5*, are no longer needed.

Figure 6-6 shows in greatly simplified form one end of a totally digital telephone system (telephone-to-telephone). The telephone set itself converts speech signals to digital signals. Separate circuits for the transmit and receive signals carry the signals from each telephone set to a multiplexer switch. In this example, the signals from two sets are labeled A and B, respectively. The switch multiplexes these and other signals into one continuous bit stream. (This process is explained later.) The bit stream is transmitted in serial form to a demultiplexer switch at the

When the transmission and the switching systems are both digital, hybrids, echo suppressors, and certain other analog devices are no longer needed.

**Figure 6-5.
Two-wire to Four-wire
Interface for Analog
Toll Trunk**

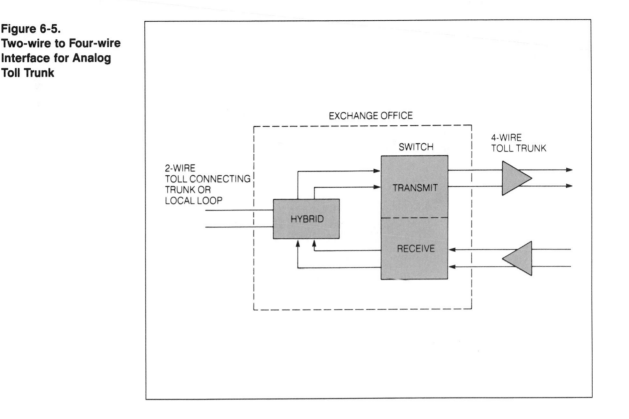

destination. This is shown as the receive function in *Figure 6-6*. The switch separates the individual signals from the bit stream and routes them to the proper telephone set.

As compared to analog multiplexing, conversion to digital multiplexing is less expensive.

The elimination of hybrids, echo suppressors, and other analog devices decreases costs and greatly increases the transmission quality for long-distance calls. In fact, if the network were digital from end-to-end, toll calls would have the same quality as local calls regardless of the distance.

Easy to Multiplex

Digital signals are easy to multiplex and the same simple low-cost digital logic circuits mentioned before can be used. The filters necessary to separate channels are much simpler than for analog multiplexing, and the need for individual wire pairs to carry the transmission is reduced significantly by multiplexing.

Status signaling is accomplished by assigning one signaling bit per channel to indicate telephone status.

Easier Supervision

Signaling for channel supervision and dialing is made vastly simpler and cheaper. Such signaling is inherently digital; for example, the on-hook/off-hook signal and the dial pulse train are binary 0 and 1 level

Figure 6-6.
Digital Transmission
Using Multiplexing
Technique

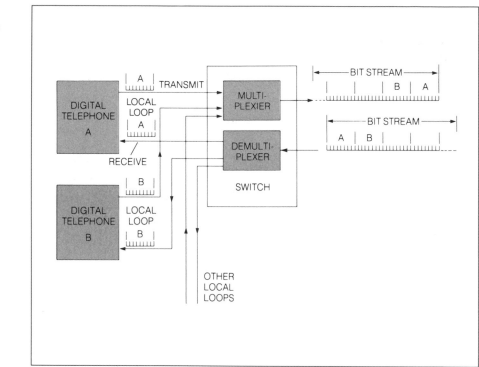

signals and can be represented with simply another bit in the transmitted digital stream as shown in *Figure 6-7.*

Better Performance in the Presence of Noise

As compared to analog, digital signals can operate in the presence of much higher noise levels because they can be more easily detected and processed from a noisy signal.

Binary signals, because they are represented by pulses of well defined and uniform shape, are easy to reconstruct even if badly distorted by noise. The process is shown in *Figure 6-8* at various points in a digital system. *Figure 6-8a* shows a *bipolar* bit stream. This slightly different format is called bipolar because the 1s of the code alternate as positive or negative levels around the 0 level. The format of *Figure 6-1a* is called a neutral bit stream because the 1s are all positive with respect to the 0 level. In communications terms, the technique of *Figure 6-8* gives the ability to communicate with low signal-to-noise ratios. The analog network needs signal-to-noise ratios of between 40 and 50 dB to provide satisfactory speech quality. This is because the noise is amplified along with the voice signal, so only a small amount of noise can be tolerated. Digital systems provide error-free performance with signal-to-noise ratios as low as 15 to 25 dB with no amplification of noise.

**Figure 6-7.
Signaling in the Digital
Bit Stream**

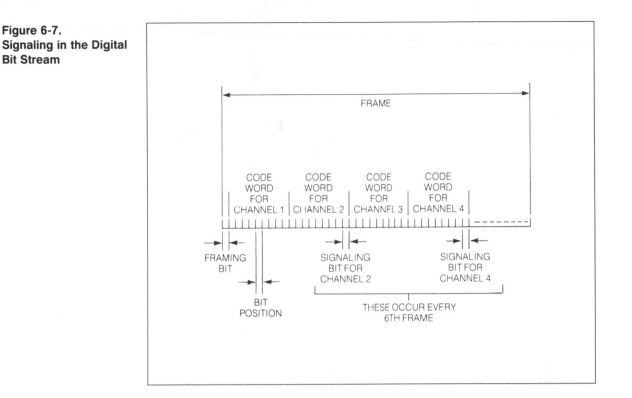

At the repeater amplifiers (*Figure 6-9*) each pulse is regenerated just as it was transmitted, which eliminates any noise added along the transmission path. Since the noise is not mixed in and amplified with the signal, the speech at the end of a 2,000 mile circuit is as clear and quiet as if the circuit were only two miles long. The closer the repeaters are spaced, the lower the probability that a digital pulse will be destroyed by noise, and the higher will be the signal-to-noise ratio of the circuit. Thus, the end-to-end error rate can be made as small as desired by proper spacing of the regenerative repeaters.

Reduced Crosstalk

Digital transmission
techniques are almost
impervious to common
line crosstalk.

Digital signals are also highly resistant to crosstalk. Crosstalk is most evident and most annoying when the two parties to a call are not talking, and can hear and understand a conversation on another circuit. Aside from being annoying, such instances breach the privacy of the parties on the other circuit. When crosstalk does appear on digital systems, it is heard as random, unintelligible noise rather than understandable speech.

**Figure 6-8.
Binary Signals in the
Presence of Noise**

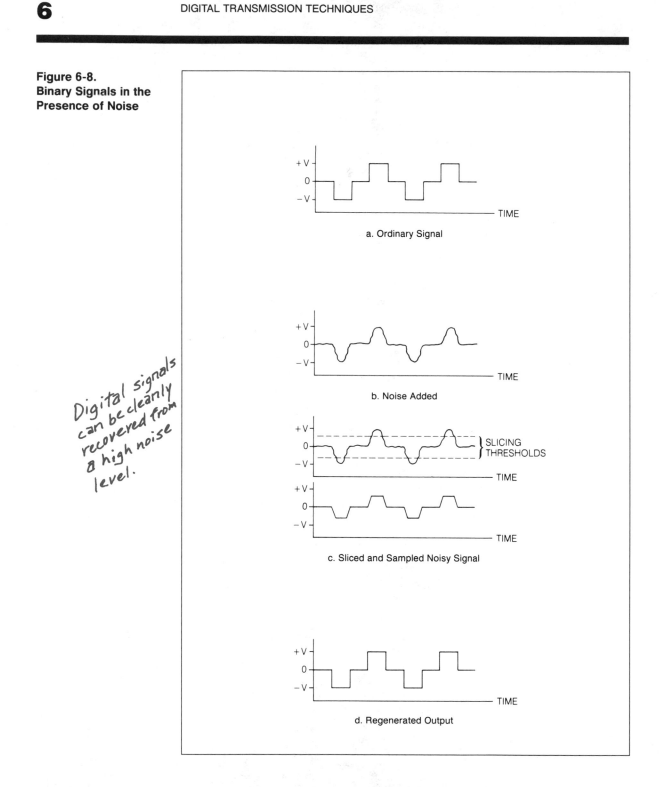

a. Ordinary Signal

b. Noise Added

SLICING
THRESHOLDS

c. Sliced and Sampled Noisy Signal

d. Regenerated Output

*Digital signals
can be clearly
recovered from
a high noise
level.*

**Figure 6-9.
Repeater
Reconstruction of
Noisy Signals**

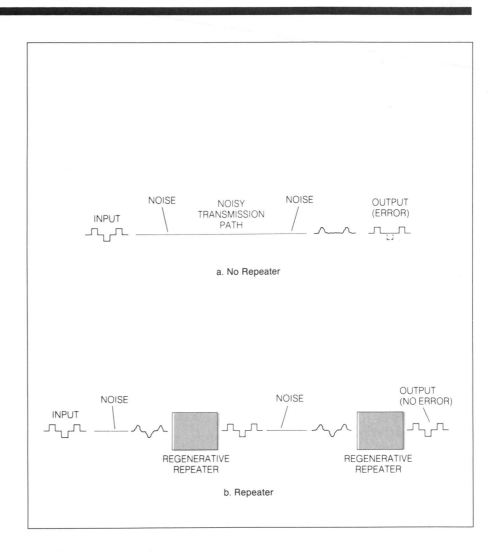

a. No Repeater

b. Repeater

Signals Can Be Mixed

The same digital equipment can be used to process all digital signal sources, whether they be data or voice, regardless of the source or destination.

Digital transmission channels readily handle digital signals from sources other than speech. *Figure 6-10* shows a string of bits from a data channel being mixed with speech that has been digitized with an A/D converter.

Since signals from all sources are digitized or are inherently digital, they all have the same form; thus, no special electronic tricks must be performed to keep them apart or to provide special quality channels. For the same reason, it is easy to monitor, and if necessary correct, the performance of digital channels as is done by the repeater amplifier. The monitoring circuit need only discriminate between the presence or absence of a pulse, without regard to the source or destination form of the information.

**Figure 6-10.
Mixing Speech and
Nonspeech Signals in
a Digital Network**

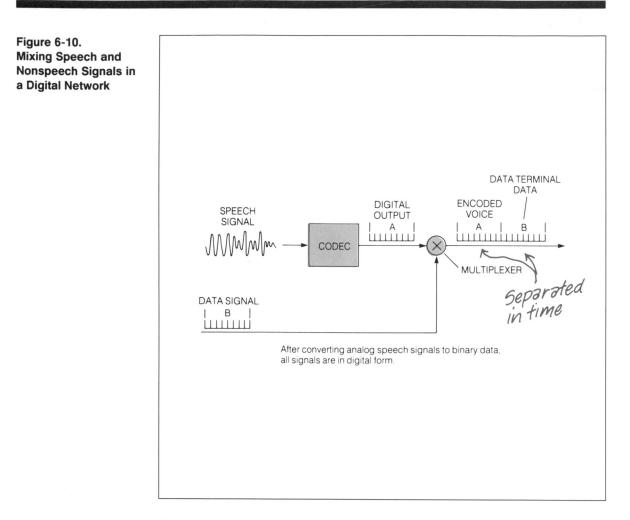

After converting analog speech signals to binary data,
all signals are in digital form.

Disadvantages

As with most real-life situations, digital transmission of speech
signals is not without drawbacks. Fortunately, most are related to the
necessity for interfacing with the existing analog network, rather than
being due to the technique or the system itself. The interface problems
will become fewer as more of the network is converted to digital
transmission and switching.

Extensive A to D Conversion

When a digital link must be connected to an analog network
element, there is a requirement for analog-to-digital or digital-to-analog
conversion. As shown in *Figure 6-11,* two such points occur at every
analog office in the network. Until such time as all of these analog

*Because of the neces-
sity of interfacing with
existing analog equip-
ment, additional con-
version equipment
(A/D and D/A) is
required.*

switching and transmission facilities are replaced, the full benefits of digital technology will not become universally available.

**Figure 6-11.
A/D and D/A
Conversions**

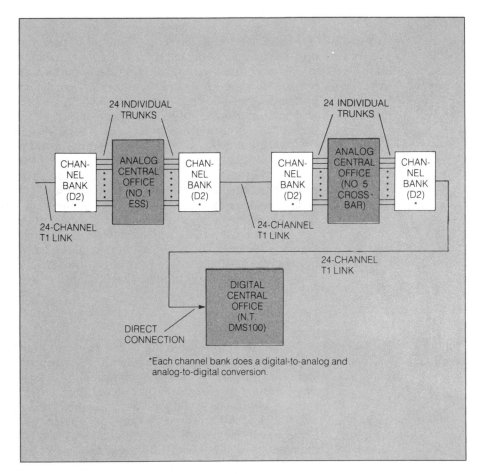

*Each channel bank does a digital-to-analog and analog-to-digital conversion.

Transient Signal Environment

Interfacing digital hardware with existing analog equipment creates some problems. Digital hardware operates on low voltages and currents, and is quite susceptible to the signal spikes generated by the existing high voltage equipment.

As mentioned earlier, there are advantages in using the standard digital integrated circuits like those used in computers. However, using such circuits in the existing telephone system also leads to a disadvantage. The integrated circuits operate on low voltages and currents. They cannot tolerate high voltages and their performance degrades if the temperature goes beyond certain limits. Most of the existing telephone network (especially the central office and subscriber loop) was designed in the era of high-current relays and electromechanical switches. As shown in Chapter 4, such circuit conditions produce high-voltage transients. In addition, such systems did not require much temperature control; consequently, temperature variations are quite extreme. The incompatibility between the two design requirements tends to be a drawback to adding digital systems to the existing network.

Maintaining Analog System Interface

The system that allows digital and analog interfacing is known by the acronym "BORSCHT."

Throughout the present telephone system, the direct current to operate the telephone and the alternating current to actuate the ringer must be separated from the digital logic circuits in a digital system that carries the encoded speech signals. In addition, other signal conditioning and protection circuits must be provided. The interface circuits that provide this separation perform functions that are known in the telephone industry as BORSCHT. The acronym stands for the functions of *B*attery feed, *O*vervoltage protection, *R*inging, *S*ignaling, *C*oding, *H*ybrid, and *T*esting.

Digital and analog interfacing is provided for each subscriber on a card called a SLIC.

As shown in *Figure 6-12,* all the BORSCHT functions usually are contained on one plug-in printed circuit card called a *S*ubscriber *L*ine *I*nterface *C*ard (SLIC) which is located in the digital switching facility. The SLIC provides the interface circuits to connect one subscriber line to the digital switch. These circuits will be covered in more detail in Chapter 7.

All of these functions except Coding and Testing have been described in previous chapters, but the point to be made here is that if the network were digital all the way out to the telephone set (end-to-end), a great deal of money would be saved by not having to provide all of the BORSCHT functions.

Increased Bandwidth

Digital transmission requires about eight times the bandwidth as analog, because of the fast circuit rise time. This adds to the amplifier and transmission line costs.

The most noticeable disadvantage that is directly associated with digital systems is the additional bandwidth necessary to carry the digital signal as opposed to its analog counterpart. A standard T1 transmission link (*Table 1-5*) carrying a DS-1 signal transmits 24 voice channels of about 4 kHz each. The digital transmission rate on the link is 1.544 Mbps, and the bandwidth required is about 772 kHz. Since only 96 kHz would be required to carry 24 analog channels (4 kHz $\times$ 24 channels), about eight times as much bandwidth is required to carry the 24 channels digitally (772 kHz $\div$ 96 kHz = 8.04). The extra bandwidth is effectively traded for the lower signal-to-noise ratios.

Accurate and Synchronized Timing

Timing errors, i.e., anything less than nearly perfect frequency and phase synchronization between the data stream and the clock, will cause data errors to be generated.

A technique for monitoring the state of just one bit in a group of bits was described at the beginning of the chapter. Since it is only possible to detect the presence or absence of a bit and to distinguish the meaning of one bit from another by their relationship in time, the element of timing is critical in digital transmission and switching systems. *Figure 6-13* shows how timing errors may cause bit errors in the signal. For a given switching system, the timing for all internal data transfers and transmissions is supplied by that system's internal clock. If this signal is received by a receiver running on its own clock, and that clock is running slow (as indicated in *Figure 6-13*) or fast with respect to the transmitter clock,

**Figure 6-12.
BORSCHT Functions**

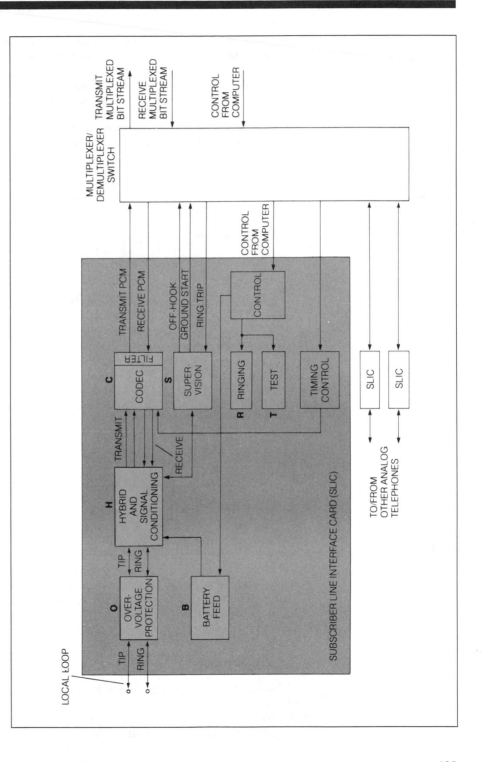

some bits will be lost as shown in *Figure 6-13*. This occurs because the signal level present at the time of transition from 0 to 1 and 1 to 0 of the clock is the level used to recover the signal. To eliminate this problem, some receivers use a method to produce a clock signal from the digital signal itself thus, synchronization is maintained. If several digital circuits from several sources terminate in a single receiving switch (which is normally the case), provision must be made in the receiving circuits to accommodate the small differences in sending rate of the various sources.

Figure 6-13.
Effect of Timing Errors
on Signal Recovery

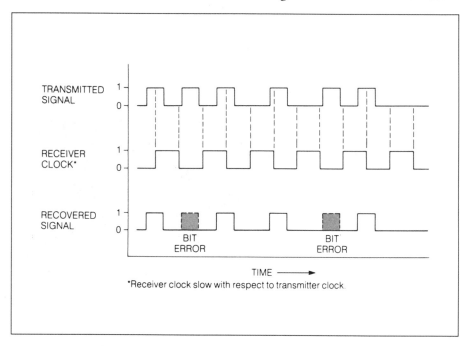

The key to digital transmission is having synchronization of the data stream and the clock. Large differences can be avoided by using a master clock source.

Small differences in transmission rates can be compensated for by using common low-cost ICs, but compensation for large differences would require an extensive amount of sophisticated circuitry. Large differences can be avoided by synchronizing all of the switches in a digital network, usually from a master or central timing source. This is the scheme used in the public switched network shown in *Figure 6-14*. It contains several different types of switching systems, but the master clock times all of the switching circuits in each of the systems. Now, with these advantages and disadvantages in mind, let's look at the specifics of digital transmission systems.

WAVEFORM CODERS

Recall that speech signals are analog signals that vary continuously with time at frequencies over the voice band from 300 to 3,000 hertz. In order to transmit over a digital system, the analog signal must be digitized by an analog to digital converter. After the conversion, speech signals can

**Figure 6-14.
Timing
Synchronization Using
a Master Timing
Source**

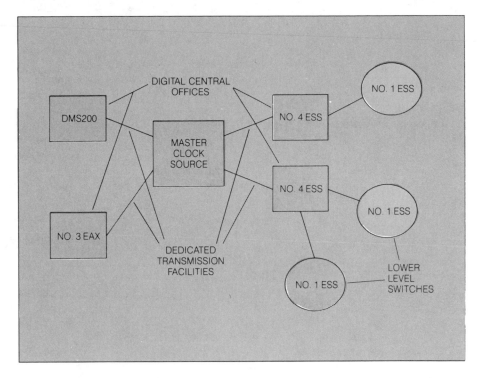

be carried very efficiently in the form of digital pulses because the pulses necessary to represent the speech can be extremely short. As a result, the pulses can be shortened in time so that many of them can be sent in the same length of time as is occupied by the original speech. This technique is called time division multiplexing. It will be covered later in this chapter, but first, let's find out how the pulses representing the analog speech are produced.

Sampling an Analog Wave

To convert analog signal into digital, it must be sampled at a rate of at least twice that of the highest frequency expected in the analog signal.

If a person speaks into a microphone and the electrical signal generated by the input sound wave is displayed on an oscilloscope with the time base set fast enough so one-half of a cycle of the signal variation is easy to see, the waveform will appear as shown in *Figure 6-15*. The sampling times are shown superimposed on the waveform. It is apparent that the amplitude of the signal does not change very much in the short interval of time between samples. Thus, a sample of the signal at any instant in time is a close representation of the signal for a short period of time on either side of the sample point. In fact, it has been proven that if the signal is sampled at a rate which is greater than twice the highest frequency component in the signal, the samples will contain all of the information contained in the original signal. This fundamental discovery in sampling theory was made by Harry Nyquist in 1933, and is appropriately known as the *Nyquist criterion*. It is described by the equation:

$$f_s \geq 2 \text{ BW}$$

where,

f_s is the sampling frequency,

BW is the bandwidth of the input signal.

Figure 6-15.
Sampling of an Analog
Waveform

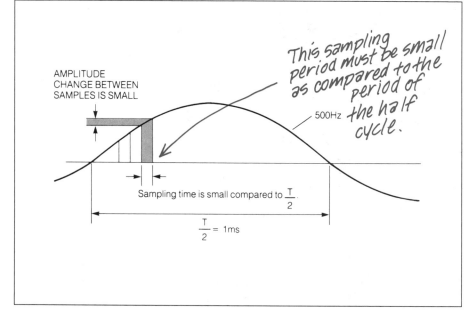

In the case of telephone speech signals, the voice channel bandwidth has been set at 4,000 Hz. Using the Nyquist criterion the sampling rate is:

$$f_s \geq 2 \times 4,000$$

which results in a minimum sampling rate of 8,000 samples per second.

Pulse Amplitude Modulation

Pulse amplitude modulation (PAM) uses a carrier of pulses that occur at a constant frequency. The amplitude of each of the corresponding pulses is made equal to the amplitude of the input signal at a corresponding sampling point.

The signal produced by this sampling consists of constant frequency pulses whose amplitude is equal to the amplitude of the sampled signal at the instant of sampling; thus, the pulses are amplitude modulated as shown in *Figure 6-16* and the process is called *Pulse Amplitude Modulation* (PAM). The sampled pulses can be sent on a digital channel and, when put through an appropriate filter at the output end, will reproduce the input signal as shown. However, there is a potential problem with this technique. Since the information is contained in the amplitude of the pulses, the amplitude relation of the pulses must be maintained. Any distortion or noise introduced in the transmission process cannot be taken out at the receiver, since the receiver has no way of knowing whether any particular sample value is distorted or not. In addition, another problem arises because the pulses tend to spread out during transmission and interfere with each other, making it more difficult to recover the original signal. Because of these problems, PAM is not normally used for transmission over distances greater than a few feet.

Figure 6-16.
Pulse Amplitude
Modulated
Transmission

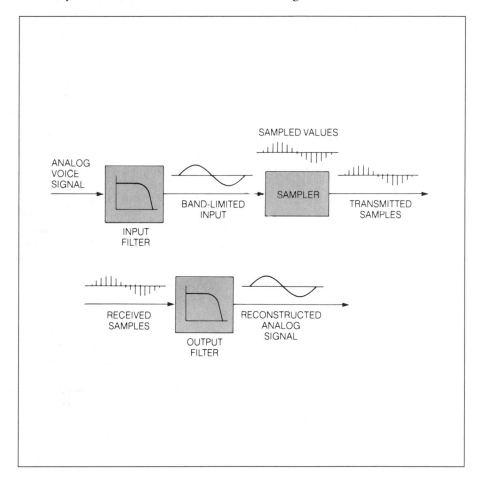

Pulse Code Modulation

Pulse code modulation (PCM) is the sampling of points along an analog wave and quantifying each sample into a coded set of binary digits.

To avoid the problems of PAM, the information contained in the amplitude of the signal sample is converted to a number. This is called quantization. The number then is coded into bits for transmission. Each of the bits in the code set have the same 1 level and same 0 level. The information is contained in the coded set of binary digits, not in the amplitude; therefore, the amplitude of the pulses can vary without affecting the information.

Quantization

The quantizer circuit converts the fixed or changing analog signal into an equivalent fixed or changing binary number.

The way a number is assigned to a particular sample is shown in *Figure 6-17*. A circuit called a quantizer takes in the sample of the analog signal and produces an equivalent number. Threshold levels are established and numbers are assigned to the samples as their amplitudes fall within the bands formed by the threshold limits. The assigned number in most cases is an approximation rather than a true value because the true value would require many more bits in the binary code. The binary code has a set number of bits which limits the unique numbers that can be assigned; therefore, the closest number to the true value is selected from the available limited set of numbers and is used to represent the sample value.

The quantization error causes background noise in the receiver—a hissing sound. Narrower sampling bands would quiet the noise, but would require a larger bit size (word), and more bandwidth.

This approximation causes an error which is the difference between the approximate number and the true sample value shown as X in *Figure 6-17*. This quantization error adds noise to the signal, called quantization noise, which is heard on the telephone as hissing. Quantization noise can be reduced by making the threshold bands narrower. This effectively provides more intervals or numbers that can be assigned over the maximum amplitude range; thus it makes the difference between the numbers smaller to reduce the quantization error. However, providing more intervals requires more bits in the binary code; therefore, more bandwidth is required. There is a tradeoff between small quantizing intervals (higher bandwidth, lower noise), and fewer intervals (lower bandwidth, higher noise).

Idle channel noise, which is caused by a low level analog signal or no signal, is effectively squelched by assigning a lower limit at a predetermined minimum above zero.

An effect called *idle channel noise* is produced in some quantizers. This occurs at small signal levels when the quantization noise is greater than the signal. This effect is particularly noticeable because there is no other signal on the channel at the time to cover up the noise. As shown in *Figure 6-18*, proper design of the quantizer results in a signal level less than the maximum value of the first quantization interval (in this case "1") being assigned a value of zero. Only the amplitudes above the first sampling threshold produce an output. This technique effectively provides an automatic squelch for small signal values and reduces the idle channel noise considerably.

Coding

The encoder (or coder) translates the quantized signal into a digital code for the numbering scheme chosen. The decoder translates the numbering scheme into the analog equivalent.

Once the analog signal sample has been quantized into a number, the number must be translated into a set of bits. The circuit that converts or translates the quantized signal is called an encoder, or usually just a coder. The circuit at the receiving end that performs the inverse operation (translating the bits into a number) is called a decoder. The combination of the two, which is necessary for a complete two-way system, is called a codec (from COder-DECoder). Codecs and their associated circuitry will be described in Chapter 7.

**Figure 6-17.
Quantization**

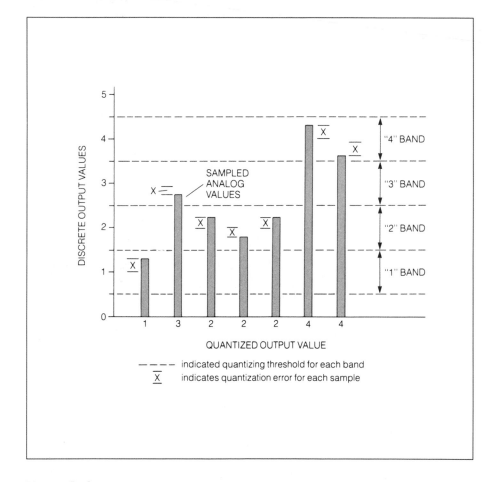

Linear Coder

Each quantization level
is assigned an individ-
ual 8-bit code. Seven of
the bits signify level
and the eighth bit sig-
nifies sign.

The simplest form of coding produces an output which is linear with
the input. A graph of input versus output is shown in *Figure 6-19*. If the
input signal value is decimal 1, the coder produces an output binary
number of 001. If the input value is decimal two, the output produced is
the binary number 010, and so on. This scheme is easy to understand. It is
used in some current business telephone systems and some of the modern
digital recording systems for high-fidelity audio recordings.

Figure 6-18.
Idle Channel Noise
Produced by
Quantization

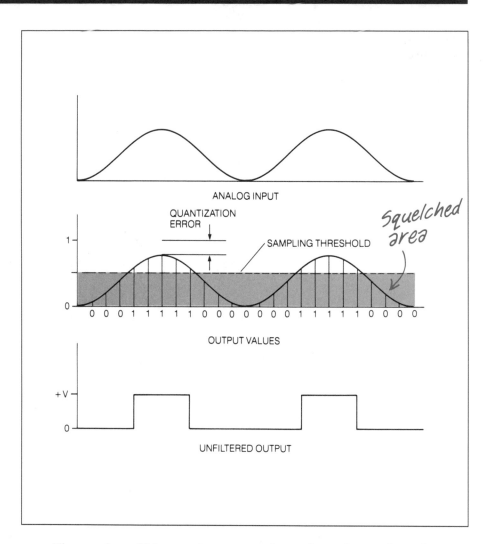

The number of bits a coder outputs depends on the number of quantization intervals in the quantizer. The number of intervals advances by powers of 2 as bits are added in the code as shown in *Table 6-2*.

It is necessary to quantize both positive and negative polarities of the signal; therefore, one of the bits of the code must be used for identifying the polarity. For this reason, the number of intervals is reduced by a power of 2; thus an 8-bit code would provide 128 intervals plus the sign bit. The number of bits in the code for a required number of intervals is:

$$n = \log_2(2 \times N)$$

where,

    n is the number of bits,
    N is the number of intervals.

Figure 6-19.
Linear Coding

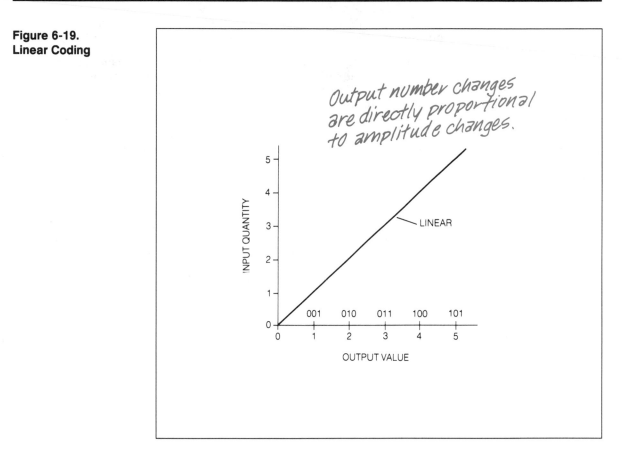

Output number changes are directly proportional to amplitude changes.

Table 6-2.
Quantizing Intervals

Number of Bits In Code	Number of Intervals
1	2
2	4
3	8
4	16
5	32
6	64
7	128
8	256

N is multiplied by 2 to gain the extra bit position for the sign bit. As an example, for 64 intervals:

$n = \log_2(2 \times 64)$
$n = \log_2(128)$
$n = 7$ bits

Companding

One important measure of the quality of a quantizer is the signal to quantizing noise ratio (SQR). The quantizer intervals are adjusted to be smaller for small signals and larger for large signals.

One of the fundamental measures of quality for a quantizer is the signal to quantizing noise ratio, or SQR. For a linear system, the SQR is the ratio of the size of the input signal to 1/4 or 0.25 the size of a quantization interval. (This value of 0.25 for the average quantizing noise is determined statistically by assuming that, over a long period of time, the coded sample inputs have a uniform distribution of levels within a particular threshold band.) This means that the SQR increases with increasing signal amplitude, so that large signals will have a higher SQR (better quality) than small signals.

In *Figure 6-20,* a small signal of amplitude 1 has an SQR of 4 while a large signal of amplitude 5 has an SQR of 20. This condition is not desirable because the small signals are more likely to occur than large signals, and the large signals tend to mask any noise present. The remedy for this condition is to adjust the size of the quantization intervals in relation to the input signal level so that the intervals are smaller for small signals and larger for large signals. This gives a nonlinear output versus

**Figure 6-20.
SQR Increases with
Signal Level in a
Linear Coder**

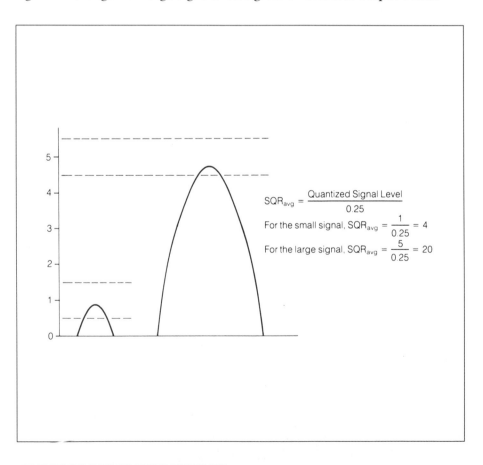

$$SQR_{avg} = \frac{\text{Quantized Signal Level}}{0.25}$$

For the small signal, $SQR_{avg} = \dfrac{1}{0.25} = 4$

For the large signal, $SQR_{avg} = \dfrac{5}{0.25} = 20$

input relationship, and results in the output being compressed with respect to the input. The corresponding curve is shown in *Figure 6-21*. Note that in this illustration, an input signal that increases in amplitude from 1/2 to 1 changes by 16 in its coded value, while a signal that changes from 1/64th to 1/32nd changes by the same amount, 16, in its coded output. Thus, a change in a small signal produces the same amount of output change as a signal 32 times as large. This system is called compression.

**Figure 6-21.
Companding Curve of
the μ-Law Compander**

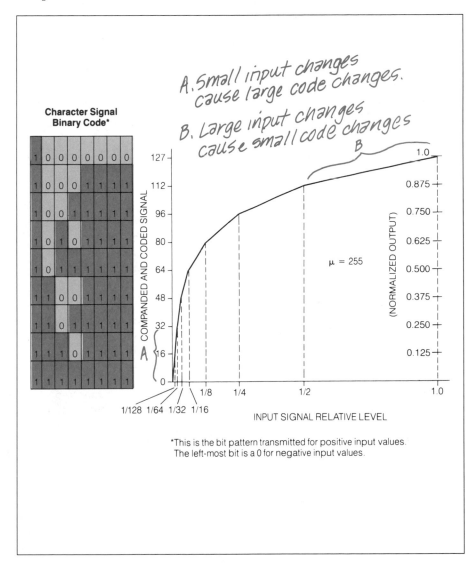

*This is the bit pattern transmitted for positive input values.
The left-most bit is a 0 for negative input values.

At the receiving end, a decoder reverses the process. The quantizer intervals are adjusted to be larger for small signals, to improve SQR performance.

At the receive end of the transmission, the decoder has a complementary expansion characteristic to restore linearity to the signal. The combination of characteristics in the codec is called a compander (for COMpressor/exPANDER). When a compander is used, the SQR is about the same across the range of input signal levels. The devices at each end of the digital transmission channels in the public switched telephone network that perform the sampling, quantization, and coding to transform speech signals to bits are called channel banks.

μ-Law Compander

Any network of digital communications systems must have a standardized scheme of doing the companding (compression and expanding). Micro-Law is the standard chosen by the Bell System.

In a large telephone system, such as the Bell System in the U.S., large numbers of digital channels are interconnected with each other and all the channel banks must use a common scheme for the analog to digital transformations and companding required. This scheme is called the "μ = 255 law companded PCM digital coding standard," usually just called μ-law. The companding circuits operate on a logarithmic curve, using the relationship:

$$F\mu(x) = sgn(x) \; \frac{\ln(1 + \mu|x|)}{\ln(1 + \mu)}$$

where,

x is the normalized input signal (between −1 and 1),

sgn(x) is the sign(+/−) of x,

μ is the compression parameter, set to 255 for the North American network,

Fμ(x) is the compressed output value.

The analog signal is encoded into an eight-bit numbering scheme, 7 bits indicate the signal level, and the eight bit indicates polarity (1 = +, 0 = −). The transmission rate is determined by the sampling rate and bit count per sample.

The encoder operates on a segmented linear approximation to the true logarithmic curve as shown in *Figure 6-21*. The encoder produces an 8-bit output; 7 bits for magnitude plus one bit for sign. The left-most bit (most significant) is the sign bit.

The sign bit is 1 for positive input values and 0 for negative input values. The remainder of the code (7 bits) indicates the absolute value of the input signal. Since the sampling rate is 8,000 samples per second, as determined earlier, the data rate for an individual voice channel when encoded using the μ-law technique is 8,000 samples per second × 8 bits per sample = 64,000 bits per second. The transmission rate on most digital facilities is much higher than this because many channels are multiplexed together.

A-Law Compander

The μ-law companding characteristic is the standard for the North American and Japanese telephone networks. For the European network, however, the companding standard is called the A-law characteristic, whose compression characteristics are defined as:

$$F(x) = \text{sgn}(x) \frac{A |x|}{1 + \ell n(A)}$$

when

$$0 \leq |x| < \frac{1}{A}$$

and

$$F(x) = \text{sgn}(x) \frac{(1 + \ell n A|x|)}{(1 + \ell n(A))}$$

when

$$\frac{1}{A} \leq |x| \leq 1$$

where,

   F(x)   is the output compressed value,
sgn(x)   is the sign(+/−) of x,
       A   is the compression parameter, set to 87.6 for the European
            network.

The European digital network is based upon the so-called A-law characteristic, which is very similar to the Micro-Law scheme. The A-Law system is quieter for small signals, but is noisier at idle channel situations.

The A-law compander also produces eight bits per input sample in the same format as the μ-law compander and a data rate of 64,000 bits per second for each channel. Its segmented companding curve is shown in *Figure 6-22*. The A-law scheme produces a slightly better signal-to-noise ratio for small signals, but the μ-law scheme has lower idle channel noise. (When plotted on the scale of *Figures 6-21* and *6-22,* the curves appear the same, but plots on a larger scale would reveal slightly different curves.)

The digitizing schemes described above take enough samples and send enough bits to encode the complete input waveform so there is exact reproduction at the destination. Therefore, they are suitable for encoding and transmitting any waveform, as long as its bandwidth is limited to what the chosen sampling rate can encode without error. The capability to accurately transmit any waveform is obtained at the cost of sending enough bits to encode the entire sample at every sample interval.

Delta Modulation

In delta modulation encoding each sample point is compared with the next and only the polarity difference is outputted in the form of a pulse train.

Another technique for encoding the sampled waveform is called Delta Modulation. This scheme, rather than sending the encoded value of the sample, sends only the polarity of the difference between one sample and the next. The basic block diagram is shown in *Figure 6-23a*. A sample of the input signal is encoded. Through a feedback path, it is decoded and compared to the input at the next sample to determine if the input signal is going + or −. In essence, the output, as shown in *Figure 6-23b,*

**Figure 6-22.
Companding Curve of
the A-Law Compander**

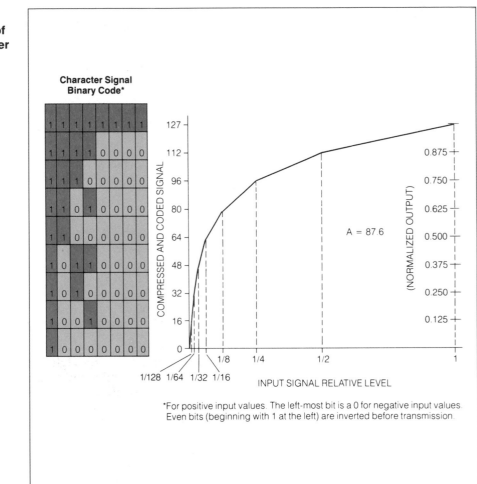

indicates only which direction the input waveform is changing, but does not indicate how fast the signal is changing. Therefore, the rate of change information must be built into the decoder or transmitted separately. In part, this problem is overcome by sampling the input at a higher rate than for the logarithmic PCM coders previously discussed. This gives better sample-to-sample correlation; thus, less error in the circuits that recover the signals. At the data rates used for standard digital signal transmission, the logarithmic PCM scheme provides slightly better quality and considerably less idle channel noise. This is because a delta modulator has no way to represent a zero output and produces a spurious signal in an otherwise quiet channel as shown in *Figure 6-24.*

**Figure 6-23.
Simple Delta
Modulation**

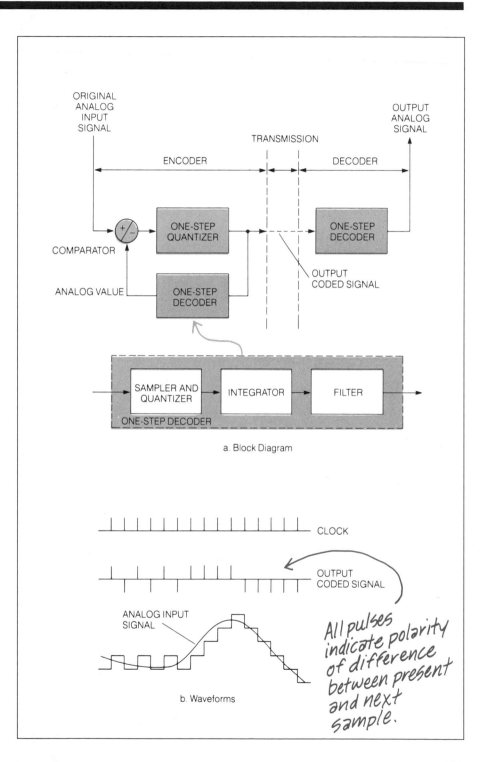

ORIGINAL
ANALOG
INPUT
SIGNAL

OUTPUT
ANALOG
SIGNAL

TRANSMISSION

ENCODER

DECODER

ONE-STEP
QUANTIZER

ONE-STEP
DECODER

COMPARATOR

OUTPUT
CODED SIGNAL

ANALOG VALUE

ONE-STEP
DECODER

SAMPLER AND
QUANTIZER

INTEGRATOR

FILTER

ONE-STEP DECODER

a. Block Diagram

CLOCK

OUTPUT
CODED SIGNAL

ANALOG INPUT
SIGNAL

*All pulses
indicate polarity
of difference
between present
and next
sample.*

b. Waveforms

Figure 6-24.
Idle Channel Noise
Generation in Delta
Modulator

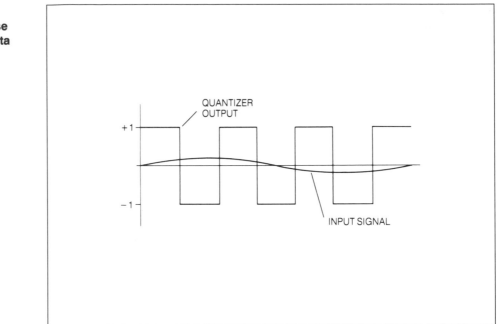

Source Coders

Another class of coders called source coders reduces the bandwidth of the digitized signal, thereby sacrificing the quality of reproduction without a significant reduction in intelligibility.

    The coders described up to this point belong to the general class of waveform coders which tries to reproduce the waveform of the input signal exactly at the output of the decoder. There is another class of coders called *source coders*, which seeks to minimize the number of bits (or the bandwidth) necessary to reproduce an input signal that is intelligible, but is not necessarily a faithful reproduction. For example, a reproduced speech signal can be understood, but may sound "hollow" or like a monotone. Source coders make use of prior knowledge of the characteristics of the source of the input signal in encoding that signal. There are several techniques currently in use, the most common of which is called Linear Predictive Coding. These are used a great deal in reproducing speech electronically. They are quite detailed and beyond the scope of this book.

TIME DIVISION MULTIPLEXING

In time division multiplexing (TDM), multiple digital voice signals are sampled in rotating sequence and put on the line. Therefore, each channel is separated in time.

    Once the input speech has been sampled, quantized, and encoded in digital form, it must be transmitted to its destination. The economics of public telephone network transmission dictate that many individual channels be multiplexed over a single large-bandwidth circuit. It is generally not economical to send only one encoded voice channel at 64,000 bits per second over a single transmission channel, although it is done in some modern PBXs using digital telephone sets. Recall from Chapter 1 (*Figure 1-15*) that for digital transmission, the method of multiplexing is to send the individual bits separated in time (rather than frequency or phase), hence the name time division multiplexing (TDM).

Figure 6-25 outlines the basic principles. In *Figure 6-25a*, a timing pulse generated from a master crystal controlled oscillator is one input to an AND logic gate. The digital code from a voice channel is the other input. When the timing pulse is present *and* an input channel pulse is present, the signal output is fed to an OR gate which reproduces at its output any AND gate output that has been turned on by the timing pulse.

The timing pulse for each channel is produced in sequence as shown in *Figure 6-25b* producing slots of time in sequence. When timing pulse number 1 is present, a time slot is created that contains the digital code for channel 1. Following it in sequence is the time slot for channel 2, created by timing pulse number 2. Channel 3 follows channel 2, channel 4 follows channel 3, and so on up to the maximum number of channels multiplexed. The cycle then repeats. In each channel time slot, the encoded symbol appears to identify the information that came in as an original analog signal.

Synchronous and Asynchronous Systems

Assignment of a place in time in a bit stream to put the bits for an individual channel can be done on a dedicated (permanently assigned) basis, or on an as-required basis. Systems that make the assignment on a dedicated basis are called synchronous systems, and involve a steady stream of uniformly spaced bits. Asynchronous systems, on the other hand, involve a start and stop bit between characters and a variable length of time. Depending on how they are designed, the asynchronous systems may be called either asynchronous TDMs, statistical multiplexers, or packet switches. In general, transmission systems in the present telephone network are synchronous, while networks designed especially for carrying digital data communications are of the asynchronous type.

Bit and Word Interleaving

In TDM, there are several different ways to perform the sampling. In word interleaving, sufficient time is allowed so that 8 bits (one byte) are taken from each channel in sequence. In bit interleaving, only sufficient time is allowed for one bit from each channel.

Each of the digital code words from each voice channel encoder contains several bits (typically eight). Four-bit words are used in the example in *Figure 6-26*, and as shown, it is possible to interleave the bits in more than one way. If the entire code word from the first channel is sent, then the word from the second, then the third, etc., the resulting set of bits is said to be word interleaved (*Figure 6-26b*). If the bits are interleaved such that the first bit is taken from each code word in sequence, then the second bit, then the third, etc., the resulting set of bits is said to bit interleaved (*Figure 6-26c*). In either case, the resulting set of bits is called a frame.

The lowest level of TDM in the public network uses word interleaving, chiefly because the source of the data (a channel bank) produces individual code words as outputs from each digitized channel. The higher-level multiplexers in the public network are bit-interleaved because the data sources are continuous bit streams from the lower-level multiplexers.

**Figure 6-25.
Basic Principles of
Time Division
Multiplexing**

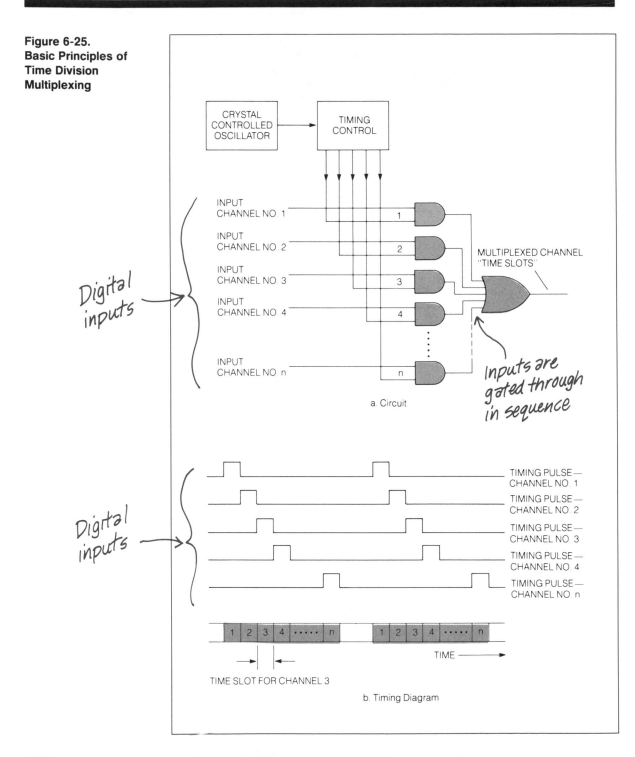

a. Circuit

b. Timing Diagram

**Figure 6-26.
Word and Bit
Interleaving**

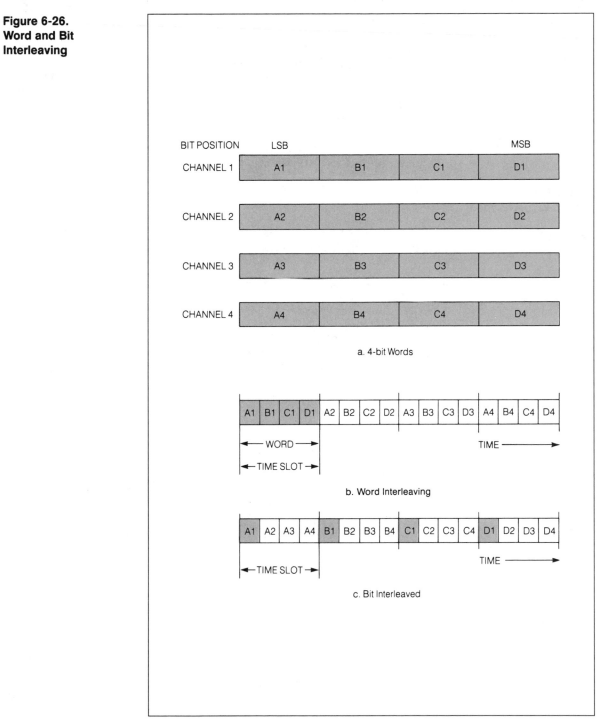

a. 4-bit Words

b. Word Interleaving

c. Bit Interleaved

Synchronization

In TDM, the boundaries of a frame are usually identified by bits or a code word.

The sending end of the multiplexed stream of bits must add framing information to the bit stream to enable the receiving end to identify the beginning of each frame. The framing information may consist of a single bit, a code word of the same length as the others in the frame, the deletion or systematic alteration of a bit in the code words, or alteration of the electrical waveform of one or more of the bits in the frame. The schemes generally used in the telephone network add either one bit or one code word (eight bits) to the data stream to identify frame boundaries. Two of the most common are shown in *Figure 6-27*.

**Figure 6-27.
Added Bit and Added
Word Framing
Methods**

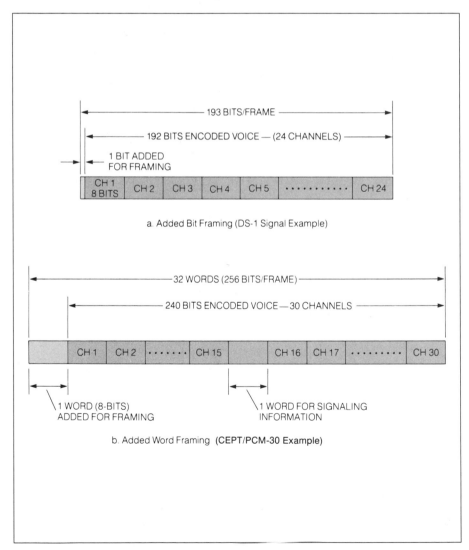

a. Added Bit Framing (DS-1 Signal Example)

b. Added Word Framing (CEPT/PCM-30 Example)

**Figure 6-27.
(Cont.)**

TI FRAME	SYNCHRONIZING BIT
1	1
2	0
3	0
4	0
5	1
6	1
7	0
8	1
9	1
10	1
11	0
12	0

24 TI CHANNELS

c. D4 Superframe Format

TI FRAME	SYNCHRONIZATION BIT
1	DATA
2	CRC
3	DATA
4	0
5	DATA
6	CRC
7	DATA
8	0
9	DATA
10	CRC
11	DATA
12	1
13	DATA
14	CRC
15	DATA
16	0
17	DATA
18	CRC
19	DATA
20	1
21	DATA
22	CRC
23	DATA
24	1

24 TI CHANNELS

d. Extended Superframe (ESF) Format

U.S. Data Rate

T1 Digital Transmission Format

Analog voice signals are generated in each local subscriber loop. These signals are received by the local exchange, where they are quantized and coded into digital information channels. In order to carry multiple channels to a destination exchange economically, information from each channel can be time division multiplexed onto a single transmission medium just as described in the preceding section.

U.S. System

Basic digital multiplexing in the United States is known as T1. It is also called the *Primary Rate Carrier* system, or simply "primary rate." Using the T1 system, 24 digitized voice channels may be multiplexed together over a 4-wire cable (2 wires for transmit and 2 wires for receive).

The format used to frame transmitted data in the T1 system is called DS-1 and is shown in *Figure 6-27a*. DS-1 partitions data into frames of 193 bits. The first bit is always interpreted as a framing synchronization bit. The 192 remaining bits represent 8-bit interleaved words from 24 channels.

Remember that a voice channel must be sampled at about an 8-kHz rate to render a clear representation of the sampled signal. This means that one 8-bit sample must be taken every 125 microseconds. Since 24 individual channels must be read for each frame, in addition to a framing bit, the system must send 193 bits in 125 microseconds. At this rate, T1 must send or receive data at $(193/125 \times 10^{-6})$ or 1,544,000 bits per second, or 1.544 megabits per second (Mbps).

European System and Data Rate

Multiplexing equipment used in Europe incorporates a different format than DS-1. The CEPT (*C*onference of *E*uropean *P*ostal and *T*elecommunications administrations) or PCM-30 format specifies a 32-word frame of 256 bits as shown in *Figure 6-27b*. The first word is an 8-bit framing word. It is followed by 8-bit data words for 15 channels. An 8-bit signalling word is then inserted, followed by 8-bit data words for the final 15 channels.

Each channel is still sampled at the 8-kHz rate, which means that an entire frame must still occur every 125 microseconds. Transmitting 256 bits in 125 microseconds yields a data rate of $(256/125 \times 10^{-6})$ 2,048,000 bits per second, or 2.048 Mbps.

T3 Digital Transmission

The T1 data rate of 1.544 Mbps is often adequate for many routes that carry a low to medium volume of traffic. On high-volume routes, however, the T1 format can be quickly stretched to its capacity.

AT&T made the first attempt at expanding its data capacity with the introduction of the DS-2 rate in the 1970s. The DS-2 format was made up of four DS-1 (T1) channels that were bit-interleaved to form a single 6.312-Mbps circuit. In order to allow for timing variations in individual DS-1 circuits, another layer of framing was added. Bit stuffing techniques were used to set the same bit rates for each DS-1 prior to interleaving.

DS-2 popularity was limited and there were few installations. Inherent problems with copper transmission media at high frequencies required the use of special cable and shielding. Development of optical fiber by the end of the 1970s opened up a whole new realm of digital capacity. The DS-2 rate was combined with developing hardware technologies and worked into a new standard known as DS-3 (T3). DS-3 is defined as seven DS-2 signals. This is also equivalent to 28 DS-1 signals which equates to 672 individual channels working at an aggregate data rate of 44.736 Mbps. Technically, DS-2 still exists, but for all practical purposes, it is an internal rate within a T3 system.

In spite of its speed, DS-3 is not quite as efficient as DS-1 or DS-2. Ninety six percent of DS-3 transmissions contain actual data, while four percent contains the checking, framing, and other "overhead" bits. This gives DS-3 an efficiency of 96%. When compared to 99% efficiency in DS-1 and 97% efficiency in DS-2, an efficiency of 96% may sound just fine, but at a data rate of 44.736 Mbps, 4% equates to 1.728 Mbps overhead—more than a single DS-1 circuit! Much of this overhead is necessary to ensure proper synchronization between DS-1, DS-2, and DS-3 formats.

D4 and Extended Superframes

The 193rd bit used to frame and align T1 transmissions can also be used to synchronize "superframe" transmission structures. This allows the T1 system to handle a much larger number of channels than the 24 channels that are synchronized within a single frame. The most common superframe structure is known as D4 (named for the D4 channel bank hardware developed by AT&T). A D4 superframe consisting of 12 T1 frames uses the framing bit to repeat a specific pattern. Framing bit patterns are interpreted by receiving terminal equipment which decodes each channel. *Figure 6-27c* shows the format for a conventional D4 superframe.

Each "frame" consists of 24 T1 channels as shown in Fig. *6-27a*, but the framing bit is varied in a set pattern that a receiving channel bank can synchronize with. This 12 bit pattern repeats every 12 frames as "100011011100." A D4 channel bank will carry 24 × 12, or 288 channels.

The Extended Superframe (ESF) introduced by AT&T provides additional signalling, diagnostics, and error detection, as well as a full 24 T1 frames to carry 24 × 24, or 576 channels. The structure of the ESF is much more complicated than the D4 superframe as *Figure 6-27d* illustrates. ESFs use the 193rd bit not only for frame synchronization as with the D4 superframe, but for a data link control channel and error detection using a Cyclic Redundancy Check (CRC) technique.

Frame synchronization bits are inserted in frames 4, 8, 12, 16, 20, and 24. These six bits are codes as "001011," which repeats for every frame. Date link bits 1, 3, 5, 7, 9, 11, 13, 15, 17, 19, 21, and 23 form a 12-bit subchannel that can be used to handle diagnostic activity without requiring a separate channel. Communication over this data link uses a standard protocol such as HDLC. Finally, a CRC enhancement reduces false framing problems that have plagued earlier D-channel bank designs. Bits 2, 6, 10, 14, 18, and 22 will cause the channel bank to reset its framing pattern if the CRC indicates an error.

Pair Gain Systems

Digital multiplexing is often used at the local level to reduce the number of wire pairs in rural areas. Some systems multiplex up to 80 subscribers on a single line.

Systems that use the techniques described to cut down on the number of wire pairs needed to carry telephone channels are sometimes called *pair gain systems*. The simplest pair gain system multiplexes only two conversations on a single wire pair, usually using frequency division techniques. More sophisticated systems such as the Subscriber Loop Multiplex system concentrates up to 80 subscribers on a single T1 carrier system. The Subscriber Loop Carrier 40 system can carry 40 subscribers on a 40-channel T1 line at 38,000 bits per second per channel. These systems are attractive in rural areas where the cost of providing individual wire pairs for each subscriber is prohibitive.

Analog and Digital Multiplexer Systems

Placing ever larger numbers of channels on a single transmission facility brought a necessity for a family or hierarchy of multiplex systems, both in the analog and digital domains. Refer back to *Table 1-5* which shows the North American analog multiplex systems and the transmission medium used for each channel. *Table 1-6* is a similar table of the North American digital multiplexer systems and the medium used. The European standard systems are shown in *Table 6-3*.

**Table 6-3.
European Multiplex
Systems**

Level	Number of VF Circuits	Multiplex Designation	Data Rate (Mbps)
1	30	Primary	2.048
2	120	M12	8.448
3	480	M23	34.368
4	1920	M34	139.264
5	7680	M45	565.148

Note that the European system is not integrally related to the U.S. system either in number of channels or data rate, making a rather complex interconnection problem at any level above that of the individual channel.

Data Under Voice

The Bell 1A-RDT system is a digital capability added to an existing analog carrier.

The Bell System has developed a special radio terminal, called the 1A-RDT, which allows the addition of a single 1.544 Mbps T1 digital channel below the lowest frequency of a 600-channel master group analog multiplex carrier system. This system was developed to implement a service offering of the Bell System called Dataphone Digital Service (DDS)®[1] for data communication. Thus, a digital capability was added to an existing analog long-haul voice carrier system.

Signaling in the PCM Word

Signaling the on-hook/off-hook state in the PCM word involves using the least significant bit (LSB) in every 6th code word. Audio quality suffers slightly.

Signaling information is carried in an analog channel by one of the several methods discussed in Chapter 1, the earliest and still most common method being that of opening and closing the dc loop. Since a PCM channel has no DC continuity, other means must be found for conveying on- and off-hook and dialing signals along with the speech. The method used in current D-2, D-3, and D-4 channel banks involves using a specified bit in a specified code word as a binary indicator of the on-hook/off-hook state of the channel. The bit used is the least significant bit in every sixth code word (*Figure 6-7*). Use of this bit which normally carries part of the speech information introduces a small amount of error or distortion into the speech signal, and lowers the SQR ratio by about 1.8 dB. A more extensive treatment of these methods is given in Chapter 8.

WHAT HAVE WE LEARNED?

1. Digital systems use information in binary form. Code words consisting of bits that have binary values 0 or 1 are used to carry information.
2. Eight-bit words are a common bit group in telephone systems. The information in the code word may be contained in the total group, in subgroups within the group or in individual bits of the group.
3. Transmission of code words in serial form is the most common method of transmitting digital signals in the telephone system.
4. Analog signals are sampled at 8,000 times per second because this is twice the maximum frequency of speech to be transmitted.
5. Analog signals are sampled, quantized and encoded into Pulse Code Modulated (PCM) binary signals. In the encoding process, the information may be compressed before transmission.
6. At the receiving end, the binary code is decoded and filtered to produce the original signal. It will be expanded if it has been compressed at the sending end.
7. The equipment that compresses and expands the binary code is called a compander.
8. There are two common types of companders; μ-law and A-law.

[1]*Dataphone Digital Service (DDS)® is a registered service mark of AT&T Co.*

9. PCM digital signals are multiplexed onto a single transmission medium by time division multiplexing (TDM).
10. Two common ways of multiplexing signals are by word interleaving or bit interleaving.
11. Multiplexed systems can be either synchronous or asynchronous.
12. Asynchronous systems are used mostly for data communications.

Quiz for Chapter 6

1. The telephone network is being converted to digital operation primarily to:
  a. carry digital computer data.
  b. reduce costs.
  c. improve speech quality.
  d. increase system capacity.

2. Digital transmission and signaling are useful because:
  a. digital logic circuits are cheaper than analog.
  b. signaling is easier.
  c. it can provide a lower signal-to-noise ratio.
  d. all of the above.

3. Digital transmission has the disadvantage of:
  a. requiring more bandwidth than analog.
  b. may not work with existing equipment.
  c. using equipment that is more susceptible to environmental extremes.
  d. all of the above.

4. The equation $fs \geq 2BW$ is called the:
  a. Shannon theorem.
  b. Nyquist criterion.
  c. Erlang law.
  d. Edison effect.

5. Sampling the analog wave produces:
  a. impulse noise.
  b. phase distortion.
  c. pulse amplitude modulation.
  d. frequency coherence.

6. Quantization assigns:
  a. voltages to digital signals.
  b. operators to incoming calls.
  c. numbers to analog samples.
  d. none of the above.

7. The simplest form of coding is:
  a. diphase.
  b. hybrid.
  c. compressed.
  d. linear.

8. A coder for a 64-interval quantizer must produce how many bits?
  a. 2
  b. 10
  c. 8
  d. 7

9. Between two quantizers, the quality of the one with the lower SQR will be:
  a. higher.
  b. lower.
  c. the same.
  d. not measurable.

10. If the code word is only 4-bits for a T1 system with a frame of 193 bits, how many channels can be transmitted?
  a. 48.
  b. 50.
  c. 6.
  d. 24.

11. In question 10, what is the bit rate in Mbps?
  a. 0.772.
  b. 3.088.
  c. 1.544.
  d. 1.024.

12. The parameter describing the degree of signal compression by the companders in the U.S. network is:
  a. sigma.
  b. μ.
  c. A.
  d. R squared.

13. In digital multiplexing systems, bit interleaving is used in:
  a. lower-level systems.
  b. higher-level systems.
  c. to interleave a code word.
  d. none of the above.
  e. all of the above.

14. Synchronous multiplexed systems have the time placement of bits:
  a. dedicated.
  b. unassigned.
  c. random.
  d. as required.

15. Asynchronous multiplexed systems are used:
  a. mostly for voice transmission.
  b. mostly for data transmission.
  c. to carry only speech information.
  d. all of the above.

Electronics in the Central Office

ABOUT THIS CHAPTER

The first five chapters concentrated on the functions of the telephone set and how the use of electronics accomplishes the functions more easily, better, or with more features. Chapter 6 started the concentration on what's beyond the telephone set. This chapter continues that emphasis, taking up specifically what's on the end of the local loop—the central office. Just as the designer of an electronic telephone set must observe the existing standards and practices, so, too, the designer of the central office equipment must be sure that the circuits will respond properly to either a modern electronic telephone set or a conventional telephone set connected to the local loop. These compatibility requirements have been discussed previously; they stem from the fact that it is usually too costly to replace all of the telephones at the time a new switching office is installed, so the new switch interface with the subscriber is designed to allow the use of existing telephones.

These standard interfaces, and how to improve their operation through electronics, is the principal thrust of this chapter. There will be, however, a glimpse of the exciting world of tomorrow's local loop, when a digital telephone connects with a digital subscriber interface and utilizes the principles of the digital switching system described in Chapter 6. In addition, much of what will be covered can be applied to specialized switching systems, such as PBXs, where the telephone set can be of any design so long as the interface between the specialized system and the central office meets the telephone company standards.

THE LOCAL LOOP

The local loop portion of the telephone network remained virtually unchanged until the late 1960s. Economics and advances in electronics technology has caused a rethinking of this portion of the telephone system.

The local loop, as discussed many times previously, connects the subscriber telephones with the local central office, and through the central office to the worldwide telephone network. The local loop operation has changed little since the invention of the telephone, though there have been many improvements in its construction. The principal change has been that twisted pair cables have largely replaced open wires strung on pole crossarms. The use of copper wires and relays as the primary system components was unchallenged for almost a hundred years, and they accounted for a major share of the capital and maintenance costs of the telephone industry. But beginning in the late 1960s, the economics of local loop design began to change. The cost of cable pairs continued to increase because of the copper, and especially because of the cost of installing them under the streets or on poles. At the same time, the cost of electronic "intelligence" was decreasing dramatically as integrated circuits became smaller, more reliable, and less expensive.

CONVENTIONAL CENTRAL OFFICE INTERFACES

All large systems are divided into modules or subsystems to simplify design, use, and maintenance. The point where each subsystem connects with another is an interface. A complete interface specification defines all mechanical, electrical, and operational rules for the inputs and outputs; ideally, it allows interconnection without requiring knowledge of the subsystem's internal operation. A large number of interfaces exist in the telephone network simply because of its size and complexity. And even more interfaces are required to adapt between old and new equipment and between the many different types of equipment that can accomplish the same function.

As we learned in Chapters 1 and 2, the central office is the place where all telephone calls are handled and first switched. If the call is to another subscriber in the same exchange, then the call is switched to that subscriber's line. If it is for a telephone in the same locality served by a different exchange, then the call must be switched to a trunk connecting the two central offices. If the call is destined for another city, it must be switched to the long-distance network via a toll-connecting trunk as illustrated in *Figure 7-1*. As it performs its switching function, the central office has two important (and different) interfaces. These interfaces are described best using telephone jargon. The local loop is commonly called

Interfaces are used to couple two telephone subsystems to each other. They contain the hardware and programming to transfer all voice, data, and power.

Subscriber calls are first handled at the central office level. Depending upon the ultimate destination, the call will be switched one or more times to different levels and carried through lineside and trunkside interfaces.

**Figure 7-1.
Typical Analog
Telephone Connection
for a Long Distance
Call**

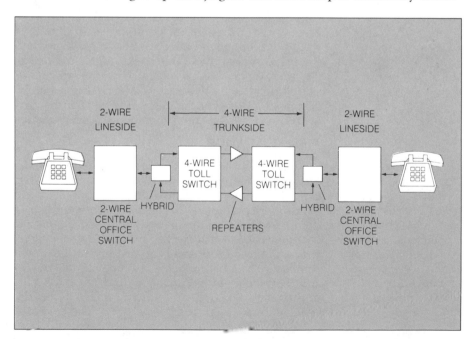

the subscriber loop or the lineside interface; and the trunk to another switching office is called the trunkside interface.

LINESIDE INTERFACE

The largest number of interfaces in the telephone network occur between the telephone set and the local office. Because this interface has evolved through the days of magneto ringers, rotary dials, and step-by-step switches, the lineside interface has been more difficult to replace with electronics and still meet the standards and characteristics that have evolved over the years. Because there are so many local loops, this is the interface that will be part of the network longer than any other.

The basic requirements of the lineside interface in the conventional telephone network are referred to as BORSCHT, which we learned in Chapter 6 means *B*attery, *O*vervoltage protection, *R*inging, *S*upervision, *C*oding, *H*ybrid, and *T*est. Let's now examine these functions of the conventional lineside or subscriber interface in more detail.

Battery Feed

The battery feed supplies the required voltage at a low resistance and high impedance.

The battery feed must provide the following for the local loop:

1. Power (typically 48-V dc) to the subscriber's telephone set.
2. The capability to allow signaling to and from the telephone set.
3. Low dc resistance.
4. High ac impedance.

Various battery feed arrangements were discussed in Chapter 3.

Overvoltage Protection

Overvoltage protection protects equipment and personnel from dangerous transient voltages due to lightning surges of up to 1,000 volts and induced voltages from, or short circuits to, utility electrical power lines. Except for higher power ratings, the protection devices for the central office are similar to those described in Chapter 3 for the telephone set.

Ringing

The central office supplies the ringing signal.

Service requests are detected by the presence or absence of current. A sensor monitors the line loop current, and detects significant changes. Two methods are used—loop start and ground start.

In Chapter 2, the telephone set ringer and ringing generators in the central office were discussed. The central office must provide the ringing signal to the subscriber telephone to alert the called telephone that a call is waiting; therefore, the central office must apply the ringing signal to the line after the switching has completed the connection. This is done normally by a relay that is energized by the switch. The ringing signal is typically 90-V rms at 20 Hz.

Supervision

Detecting service requests (when the caller goes off-hook), the dialing input, and supervising calls in progress (when the ring is answered, or when either party hangs up) are accomplished by detecting the presence or absence of current flow in the loop. This requires a sensor that can discriminate accurately, regardless of line length, between off-hook current and current as a result of noise, leakage or a small standby current for the memory in an electronic telephone. There are two common methods used for detecting a subscriber off-hook—loop start and ground start.

Loop Start

In loop start detection, a cutoff and line relays are used for sensing and logic. The line relay is energized when the switchhook contacts are closed. The line relay is disabled by the activation of the cutoff relay.

Loop start lines are used in the vast majority of local loop circuits; they signal off-hook by completing a circuit at the telephone. *Figure 7-2* illustrates a subscriber line interface using relays for sensing and logic. In the on-hook condition, neither the line relay nor the cut-off relay is operated, and the line relay battery provides power to the line. No current (except perhaps leakage current) flows because the switchhook contacts are open.

When the subscriber lifts the handset (*Figure 7-2a*) current flows from the line battery through the closed switchhook contacts and energizes the line relay. A set of the line relay contacts close to signal the switching circuits, via the line finder, that the subscriber wishes service. When the line finder seizes the line to provide dial tone, it causes the cut-off relay to operate, which disconnects the line relay as shown in *Figure 7-2b*, and extends the circuit into the switching equipment. This also disconnects the line battery so further operation is powered by another battery supply through the first selector or register of the switching system. Note that the line relay has two windings, one connected in each side of the line. The windings are balanced and wound so that voltages induced in the line are canceled; thus, only current that flows around the entire loop will cause the relay to operate.

Ground Start

In ground start, the line relay may be activated by grounding the ring side of the line at any point. This configuration is usually used with PBX equipment.

Ground-start lines are used on loops connecting PBXs to the central office, and in other situations where it is desirable to detect a line that has been selected for use (seizure of the line) instantaneously from either end of the line. Grounding the ring-side path, as shown in *Figure 7-3,* causes current to flow through one-half of the line relay which is sufficient to energize the relay. Further operation is as explained for loop start. When dial tone is detected by the PBX equipment, the ground-start contact is opened.

Dialing Supervision

Three relays are used to supervise dialing. By using different relay characteristics, the circuit detects the difference between no current due to dial pulses and no current due to the handset being on-hook.

The problem in dial pulse signaling is to detect the difference between no current due to dial pulse break intervals and no current due to on-hook. This has been accomplished in the conventional system through the use of three relays, called A, B, and C, as shown in *Figure 7-4*. The A relay is fast operating; it energizes when the tip-ring circuit is closed by the switchhook contact, then releases or energizes as the dial pulsing contacts open and close the circuit. Thus, the A relay "follows" the dial pulses and its contacts open and close in synchronization with the dial pulses to control the first selector in a step-by-step switch.

Figure 7-2.
Line Interface Relay
Operation

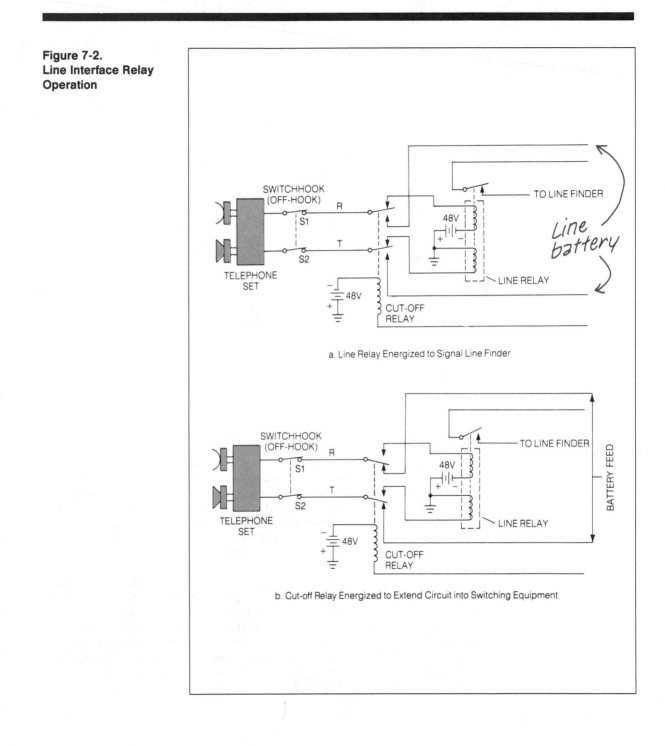

a. Line Relay Energized to Signal Line Finder

b. Cut-off Relay Energized to Extend Circuit into Switching Equipment

Figure 7-3.
Ground-Start Signaling

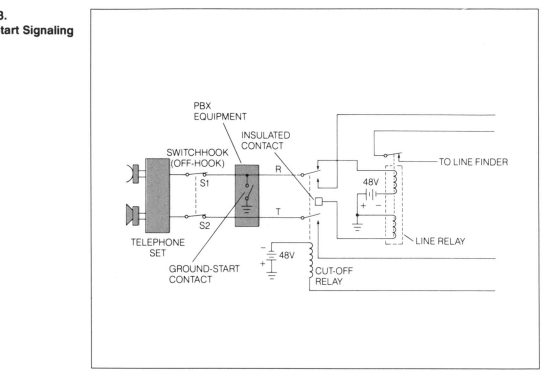

The B relay energizes when A energizes, but it is a slow-release relay. It is designed so that enough energy is stored to hold its contacts closed for a short time after energizing current is removed; therefore, its contacts remain closed while the A relay contacts are opening and closing. The B relay does not release until the A relay has been released for about 200 milliseconds; thus, the B relay is held energized as long as the subscriber is off-hook and indicates, by releasing, when a subscriber hangs up.

The C relay also is a slow-release type and is used to detect the end of the pulse train for the dial-pulse receiver. When the A and B relays are already energized, the C relay is energized when the first dial pulse interrupts the current to release the A relay. The C relay remains energized as dial pulses are generated and does not release until the A relay remains energized for an interval of about 200 milliseconds. Since the longest standard dial pulse period is less than 100 milliseconds, a 200-ms release time for the C relay is sufficient to hold the C relay energized over the longest dial pulse interval. Then, when dialing is finished and the A relay remains energized, the C relay releases which signals the switch that dialing of that digit is completed.

Figure 7-4.
Dialing Supervision

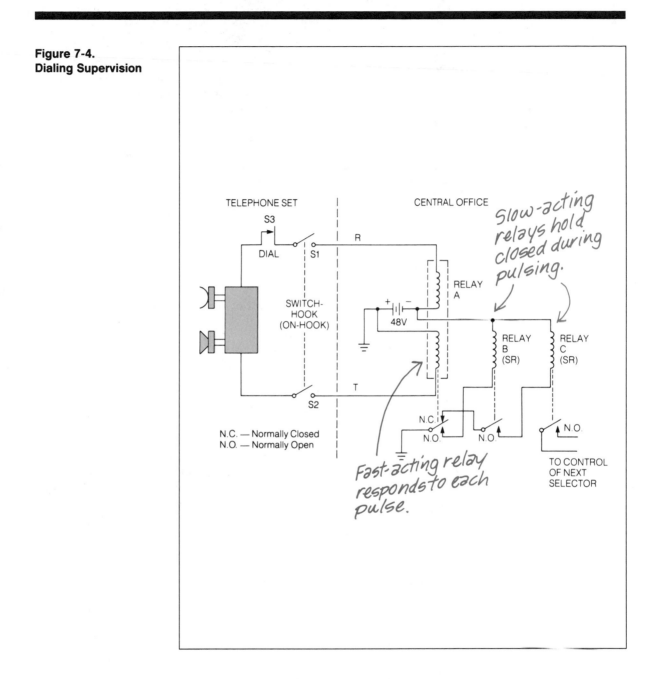

Answer Supervision

Answer supervision requires signaling the switching office that a telephone has been answered. The methods include disconnecting ring current, momentary circuit interruption, and polarity reversal.

Answer supervision involves disconnecting the ringing current (called "ring trip") when the called party answers; it also may require momentary interruption of the circuit, or reversing the polarity of the tip-ring pair. Some telephone companies use this means to indicate that the call has been "cut through" to the called party, so that charging for a toll call can begin. The possibility of polarity reversal means that the purchaser of a telephone set should insist that it be equipped with polarity protection as discussed in Chapter 3; otherwise, the set may not be able to signal remote computers or specialized carriers through the DTMF keypad. Worse yet, the set may be damaged if connected to a line where polarity reversal is used.

Coding

Codecs combine the encoder and decoder into one assembly.

The coding of the voice signal into serial digital codes that are placed into PCM time slots for digital transmission occurs in an encoder at the sending end as discussed in Chapter 6. Recovering the signal at the receiving end requires a decoder. When both encoder and decoder are combined into one integrated circuit, it is called a codec. Codecs will be discussed in more detail later in this chapter.

Hybrid

When signals are amplified because of long distance transmission, a hybrid is required to separate the send and receive signals.

The conventional local office is a two-wire switch, meaning that conversation travels in both directions over the same pair of wires. When a signal is to be transmitted over long distances, amplification is required; therefore, switches for these long-distance circuits must be four-wire as shown in *Figure 7-1*. The two-to-four-wire conversion is accomplished by the hybrid transformer as described in Chapter 2. The hybrid is on the trunkside interface of the conventional switch because local calls do not require conversion; only calls needing amplification because of distance must go through a hybrid. However, if the switch is digital, the hybrid must be part of the lineside interface.

Test

Testing requires access to the local loop circuit and to the circuits of the switching equipment to detect faults and provide maintenance. Additional relays placed in the local loop circuit provide the access required.

TRUNKSIDE INTERFACE

A trunk usually refers to the channel(s) between the equipment at two switching locations.

The word "trunk" has acquired many meanings (some conflicting) over the years. The most common use means a channel between the equipment at two switching locations. This definition still creates confusion, since for example, a "trunk" at a private branch exchange (PBX) installed in an industrial plant or in a business office ends up as a "line" at the central office, while a private circuit or "tie-line" between two PBXs is functionally identical to a trunk between two central office switches.

Trunk Circuits

A trunk circuit is the interface between the trunk and the switching system. The transmission method may be wire pairs, or multiplexed analog or digital signals. The trunk must have a battery supply, supervision signaling, and termination.

Usually a trunk is terminated in a "trunk circuit," which is the interface for transmission, supervision, and signaling between the trunk and the switching system. It is considered to be part of the trunk when measuring transmission levels. Trunks are more expensive than subscriber loops; thus, they are provided only where needed. However, because the trunk circuit is not dedicated to one customer, they usually have much higher usage than local loops. For example, a group of 30 trunks which are designed so that only 3% of the calls during peak traffic will be blocked (not connected) will average nearly 70% usage over that peak period.

Short distance trunks may be pairs of wires, while long-distance trunks are usually implemented through multiplexed analog or digital carrier systems. The trunk itself may be one-way or two-way, and either automatic or operator-handled. The trunkside interface at the central office accommodates these varying types, and provides the same sort of functions that the lineside interface provides, although with more variations and complexity.

Figure 7-5 illustrates some of the interfaces on the lineside and trunkside of AT&T's computer-controlled space-division No. 1 *Electronic Switch System* (No. 1 ESS) showing typical switching paths and some of the terminology. Connections are routed through crossbar switches to connect line-to-line or line-to-trunk. Battery, supervision, signaling, and termination all are required on trunks.

ELECTRONICS IN CENTRAL OFFICE INTERFACES

For a number of years, the No. 1 and No. 1A ESS have been the principal class-5 switches in the Bell System. New designs have produced the No. 10A RSS (Remote Switching System). It is designed to be controlled remotely using a No. 1 ESS as the main controller. The No. 10A provides essentially all central office functions, and allows the use of electronic switching in areas where it formerly was not economical to do so. The No. 10A is, in fact, another example of distributed intelligence in the telephone network.

Electronic SLIC

The No. 10A subscriber line interface circuit (SLIC) and switching electronics combines the best advantages of solid-state and more traditional electrical methods.

Our interest in the No. 10A is its *Subscriber Line Interface Circuit* (SLIC) and switching electronics. The switching portion will be discussed later in this chapter, but for now let's use the No. 10A's line interface, shown in *Figure* 7-6, to illustrate various ways of providing SLIC functions. The No. 10A SLIC is a circuit that is intermediate between a full integrated circuit and no electronics because it combines electronics with a transformer to provide:

**Figure 7-5.
Interfaces of
Computer Controlled
Switch**

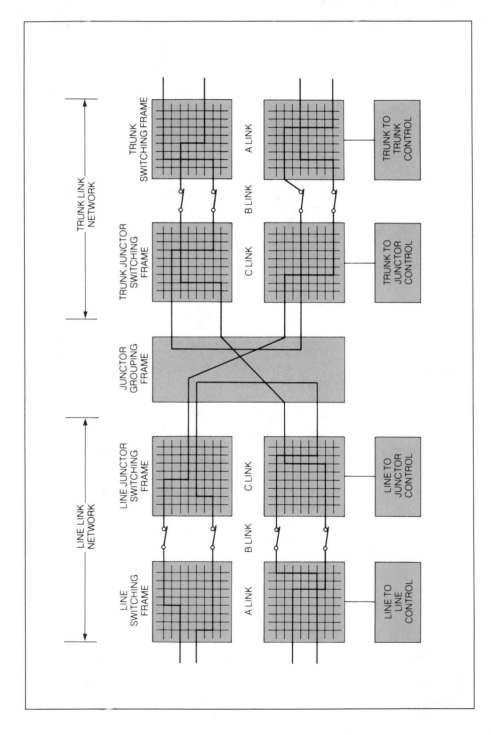

Figure 7-6.
Bell No. 10A RSS SLIC

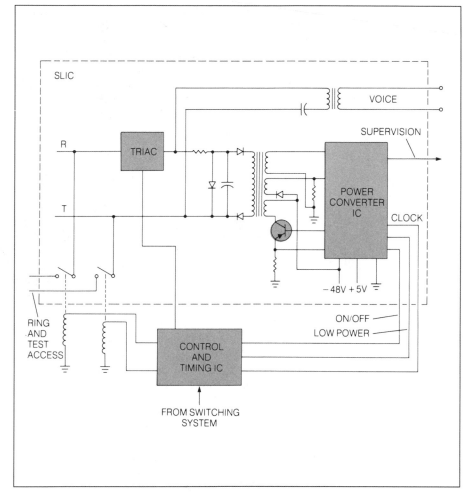

1. Low power dissipation of 650 mW per line.
2. Isolation from common-mode signals on the local loop by transformer coupling to the power converter.
3. Protection against high voltage transients by diodes and transformers because of a transformer's low frequency response.
4. Two modes of operation—high power and low power. In high-power operation, the power converter operating frequency changes from 40 Hz to 90 Hz depending on the line resistance. This varies the output voltage as a function of line length. In the low-power mode, the converter only supplies enough current to compensate for loop leakage.
5. Application of ringing voltage.
6. Access to the circuit for testing.

During ringing and loop testing, the power feed from the converter and the voice feed must be disconnected from the loop. This is accomplished by stopping the converter and allowing the disconnect Triac to open. A ringing voltage of 20 Hz is applied from a common ringing bus via reed relays. Reed relays also provide test access. These relays are controlled by the remote switching system via latches and relay drivers on a control and timing IC. This control and timing IC also controls the converter IC, participates in setting up the path through the switch, and multiplexes the supervisory signals, including ring trip.

Integrated Circuit SLIC

Integrated circuits have been introduced into SLIC functions because they can perform many BORSCHT functions, are compatible with digital applications, and are easy to program.

The design of integrated circuits to satisfy the subscriber line interface functions has been directed not only to satisfy as many of the BORSCHT functions as possible, but also to look ahead to the conversion of more and more of the telephone system to digital operation. In addition, the flexibility of programmable digital systems to change to different applications or incorporate additional performance features by changing the program, has influenced the design.

However, there are limitations. Semiconductor materials, especially in the structure of integrated circuits like the SLIC that handle a variety of functions of amplification and logic, are limited in the breakdown voltages that they can withstand. Therefore, the high voltage and voltage transient protection normally are not provided on the integrated circuit, but are taken care of by external components. Integrated circuit differential amplifiers have very good common-mode rejection, but have difficulty handling the large common-mode signals appearing on the local loop. Therefore, the transformer isolation available on the incoming local loop is maintained when using many integrated-circuit SLICs and the battery feed available at the central office is used as is, rather than using the power converter technique described before. When SLICs provide battery feed, the most common technique is to provide current drive from external transistors to make sure adequate current is supplied the subscriber's telephone set for proper operation.

The TCM4204

The TCM4204 is designed to interface a subscriber line to a digital codec. It is a self-contained package providing hybrid, supervision, end control of the ring and test functions.

An example of an integrated circuit designed to provide the subscriber line interface functions is the Texas Instruments TCM4204. As shown in *Figure 7-7*, the TCM4204 is designed to interface a subscriber line to a digital PCM switch and is intended to interface to a codec. It provides the hybrid, supervision, and control of the ring and test functions. Because it provides the hybrid 4-wire to 2-wire conversion, this function moves from the trunkside interface to the lineside interface of the central office switch and is provided for each subscriber line. The battery feed stays as it exists in the central office and transient voltage protection is provided by an external TCM4301 integrated circuit.

Figure 7-7.
Digital Line Card Block
Diagram

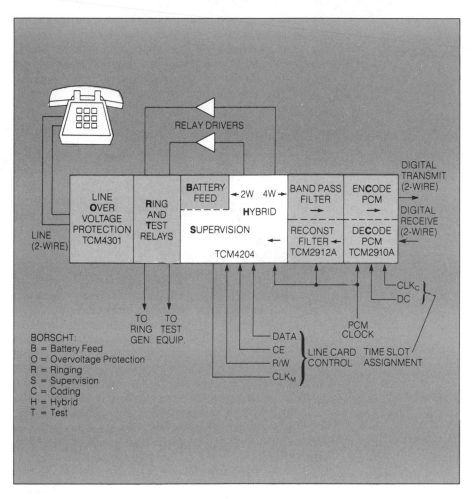

Operation

A more detailed diagram of the integrated circuit itself interfaced to the subscriber line is shown in *Figure 7-8*. Speech signals pass to the subscriber from the central office switch through the receive input (REC IN), through the receive path attenuator (REC ATTN), and through a driver that drives the line transformer in push-pull through external resistors R1 and R2. The external zener diodes, D1 and D2, provide overvoltage protection. After transformer coupling in the central office, speech signals from the subscriber are input to the TCM4204 input op-amp, A1, through the transmit attenuator (XMIT ATTN), and out on the 2-wire transmit line through the switch to the trunk circuit. The gain of the operational amplifier is set by external resistors R3 and R4.

**Figure 7-8.
Subscriber Line
Control Circuit**

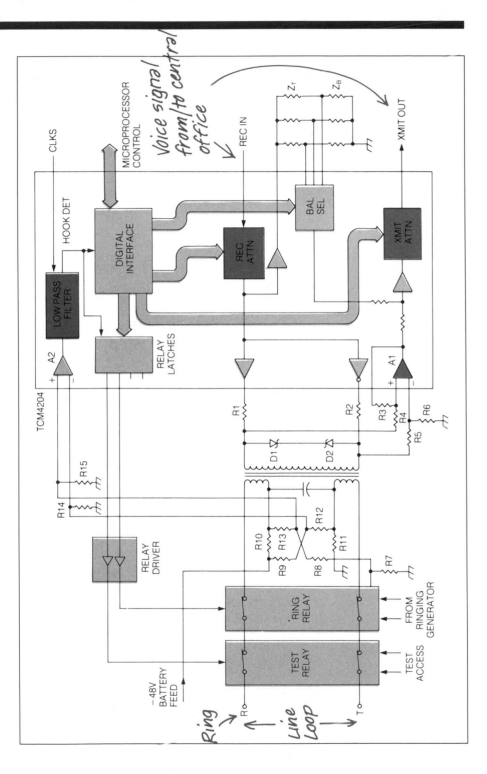

Switchhook closure (off-hook) is detected by the resistor bridge (R8 through R13) in the local loop side of the line transformer and fed through the operational amplifier, A2, to a low-pass filter and to the microprocessor control bus through the digital interface. The low-pass filter blocks the ringing signal when it is present so it will not affect the microprocessor. Dial tone is provided to the line through the REC IN input and then coupled through the receive path. DTMF dialing tones follow the transmit path and are output on XMIT OUT.

External relay drivers for the ring and test relays are controlled by on-chip latches which are timed by the microprocessor control through the digital interface. When activated, a relay driver energizes the ring relay to connect the ringing signal to the line. When the subscriber answers, off-hook current is detected by external resistor R7 to provide a ring trip signal through A2. Test access is provided by energizing the test relay in the same fashion as the ring relay. When the relays are not energized, battery feed of −48 volts is provided through the normally closed relay contacts to the subscriber loop.

Advantages

The digital line control circuit allows the remote microprocessor to control many of the operational and testing functions.

The most significant feature of such a line control circuit is the digital interface and the additional control features that it provides. It allows the external microprocessor (which is shared among many SLICs) to separately control the receive and transmit attenuators, to select the external balance network providing Z_T and Z_B, to power-up or power-down on command, and to control analog loopback for testing purposes. In addition, both ground-start and loop-start supervision are provided.

Operating conditions of the TCM4204 are held by 24 bits of memory in digital interface circuitry. External microprocessors establish and maintain control of the IC by writing to each register bit serially. Each bit can also be read serially by a microprocessor. Bits 0 and 1 are fixed conditions which can only be read, while bits 2 through 23 are temporary bits which can be written or read as needed. The specific purpose of each bit can be found in the TCM4202 data sheet.

Four discrete control lines are used to manage the SLIC: Chip Enable (CE), Data I/O (I/O), Read/Write (R/W), and Clock Input (CLKM) control timing and data transfers to or from the microprocessor. Serial data transfer takes place over the I/O pin as long as CE is logic low and R/W is either logic low or high. Each clock pulse at CLKM will advance the register's pointers to direct the next available bit on the I/O line.

Two register bits select one of four operating modes in the SLIC. Mode 0 sets the SLIC into a Standby Mode. This is very similar to a Power-Down Mode, but output status pins are signalled differently. Mode 1 initiates a Power-Down Mode. In this way, power to SLIC control circuits is maintained, but all audio circuits are disabled. Mode 2 is for normal voice operation. In this mode, all internal circuitry is powered and the SLIC will function normally. Mode 3 establishes a loopback condition which opens the normal balance circuit and forces any transmit output signal to follow the input signal at the receiver. This mode is used for loop testing.

Voice-Frequency Filters

For sampled transmission systems, the voice bandwidth must be limited at both sides of the spectrum to prevent distortion.

When sending a conversation to a called party in any sampled transmission system, it is important that no frequencies higher than half of the sample rate be input to the sampling circuit in the encoder. If input, these frequencies will not be reproduced properly, but a spurious signal will appear in the output that is the difference between the actual unwanted frequency and half of the sample rate. This phenomenon is called *foldover distortion* or *aliasing*, and avoiding it requires that the input signal be filtered before sampling. (A visual example of the phenomenon of aliasing occurs in western movies where the frame rate is too slow to properly record the fast-moving spokes of the stagecoach wheels; thus, instead of rotating rapidly forward, the wheels appear to rotate slowly backward or even stop.)

When the signal arrives from the called party, the output of the decoder must be filtered to remove the high frequencies caused by the stairstep reconstruction process that occurs in the decoder. These filtering requirements indicate the need for inexpensive, high-quality filters in digital telephone systems.

The transmit and receive filters shown in *Figure 7-7* are basically low-pass filters which reject frequencies above 3 kHz. However, the transmit filter must also attenuate any 60-Hz input component (which might be induced into the local loop from nearby power lines); thus, it is actually a bandpass filter with a passband from 300 to 3,000 Hz. As with most filters, the performance specifications are in terms of passband (300 to 3,000 Hz) and stopband (less than 300 Hz, greater than 3,000 Hz) attenuation. However, in this case, the ripple specification (variations in response across the passband) is also very important because of problems with transmission quality when call routing results in several filters being placed in series, which is likely to happen in PCM transmission systems. The stopband requirements prevent aliasing of frequencies above 4 kHz.

The filters also contribute to crosstalk and idle channel noise. If more than one filter is put in a single monolithic device, care must be exercised to provide a large crosstalk attenuation between them to ensure that they do not interact. Crosstalk coupling through a common power supply also must be avoided.

The TCM2912C

A block diagram of a typical single-chip voice-band filter, the Texas Instruments TCM2912C, is shown in *Figure 7-9*. The diagram is divided into three sections. The transmit section has third-order high-pass and sixth-order low-pass filters to provide bandpass filtering to eliminate unwanted switching and low-frequency noise. The receive section furnishes sinx/x correction for the codec and eliminates high-frequency switching signals. The TCM2912C is designed to implement the transmit and receive passband filters for PCM trunks or line terminations. It uses switched capacitor techniques and is fabricated using NMOS technology. Transmit gain can be adjusted by the ratio of resistors R1 and R2. If high impedance electronic hybrids are used, the receive output (HIGH IMPEDANCE REC ANLG OUT) can drive the hybrid directly. If low impedance coupling is required, the on-chip power amplifier can be connected as shown in *Figure 7-9*. R3 and R4 are adjusted for required gain and R5 is for impedance matching. The third section of the diagram contains clock generators, voltage regulators for receive and transmit filters, and a substrate decoupler to reduce crosstalk.

Codecs and Combination Circuits

Figure 7-7 showed the block diagram of a digital line card. It showed the encoder and decoder combined into the codec (TCM2910A) and the codec interfacing with the TCM2912C. The TCM2912C contains both the transmit and the receive path filters. The subscriber line control circuit, TCM4204, provided the required 4-wire to 2-wire conversion.

The arrangement of *Figure 7-7* is on a per-line or per-channel basis. There is also an arrangement whereby one codec is shared by a number of lines by using multiplexing techniques. This saves on equipment costs because fewer codecs are required; however, some problems exist when using shared codecs. The first is reliability or downtime. If many lines are multiplexed through one codec, many lines go down if one codec goes down. The second problem is that the analog time division multiplexing is much more difficult and less flexible than digital multiplexing. Also, shared codecs are more difficult to design in single-chip integrated circuit form due to the problems of crosstalk between the channels and obtaining the speed performance required.

Because of these problems and the potential volume (there are some 600 million subscriber telephone lines) which should reduce cost, most of the development has been directed to individual codecs per line. (The projected cost reduction follows the standard learning curve design philosophy where cost decreases as volume increases.)

The main applications of the codec in telephone systems are in central offices, channel banks, private automatic branch exchanges (PABX), and digital telephones. There may be some other nontelephone applications, but they are low-volume usage.

Operation

Virtually all codecs use the successive approximation technique illustrated in *Figure 7-10*. In the transmit direction, the two-wire signal from the subscriber is sampled in the sample-and-hold amplifier and encoded using the voltage comparator, the companding digital-to-analog converter, and the successive approximation register. The resulting compressed binary data are loaded into the data buffer and shifted out into

To save on some equipment costs, many lines may be multiplexed onto a single codec. There is a trade-off, however. Many telephone circuits would then fail if one codec failed. Also, analog time division multiplexing is tricky, and there are crosstalk problems. The prevailing wisdom is that individual codecs for each line is the best method to use.

The basic codec configuration uses digital techniques to perform companding, A/D and D/A conversion, etc. all under clock control.

Figure 7-9.
PCM Line Filter

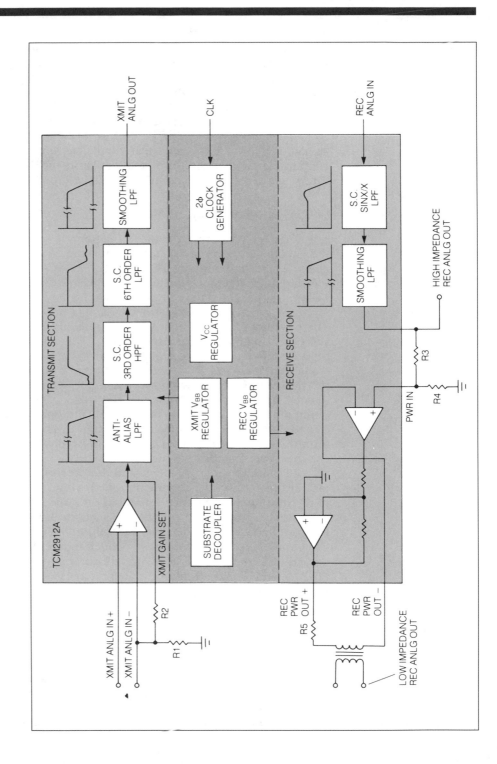

the proper PCM time slot under control of the transmit data clock. In the receive mode, the digital information comes from the line to the expanding DAC via another buffer, and is timed into the buffer by the receive data clock. The companding DAC, acting like a A/D converter, produces a voltage which is held in the receive sample-and-hold amplifier, and then is routed through the reconstruction low-pass filter for smoothing.

Figure 7-10.
Per-Channel Codec
(Source: Paul R. Gray and David Messerschmitt, "Integrated Circuits for Local Digital Switching Line Interfaces", IEEE Communications, Vol. 18, No. 3, 1980, pp. 12–23, Copyright© 1980 IEEE)

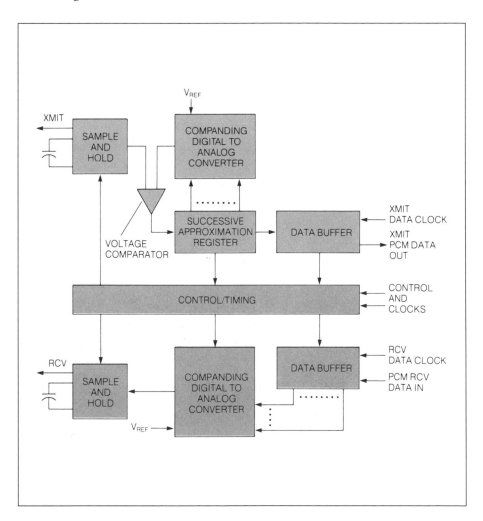

The companding coding was discussed in Chapter 6; *Figure 7-11* adds some more detail. The positive signal μ-law curve of normalized input versus output is used as an example because it is the most common in the U.S. The curve is implemented in the codec by a segmented linear approximation.

Figure 7-11.
μ-law Companding
Code Detail

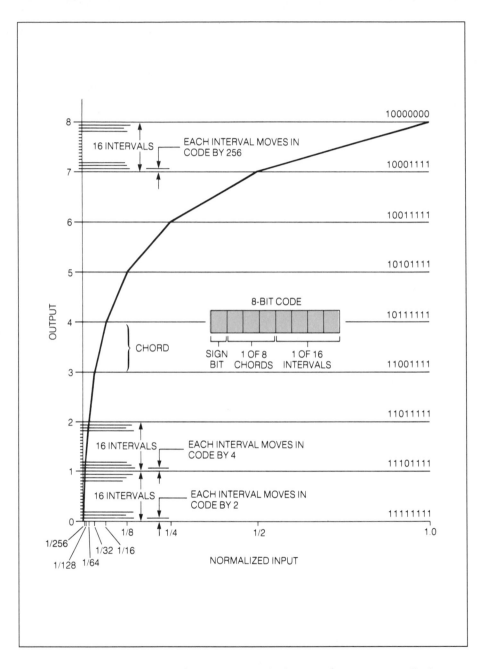

The total output signal range is divided into 16 segments called chords—8 for positive signals and 8 for negative signals. Each chord is divided into 16 intervals. Each interval in a particular chord moves in output code by the same number of bits, but as the output increases in

To achieve companding coding, the Micro-Law curve for the digitalizing of the signal by the codec is used. Each interval increase in signal level results in an increase in the 8-bit output code. This stepped increase is linear within each chord, but the size of the steps increase logarithmically for each new chord.

magnitude and spans a higher number chord, the coding produces a larger number per interval. As shown for the second output segment (chord 2), the output code increases by 4 for every interval, while for the eighth output segment (chord 8), the output code jumps 256 for every interval. The companding code internal to the codec is made up of 8 bits as shown in *Figure 7-11*. One bit is for the sign (+ or −), three bits are to identify the chord, and four bits are to identify the interval.

The TCM2910A

The Texas Instruments TCM2910A is a complete µ-law companding codec. It is designed to operate as a PCM codec which can provide all functions necessary to interface a 4-wire telephone circuit with a standard TDM digital transmission system such as T1. Each internal function of the TCM2910A can be found in the block diagram of *Figure 7-12*. It is fully compatible with the TCM2912C filter IC.

Figure 7-12. TCM2910A Block Diagram. *(Courtesy Texas Instruments, Inc.)*

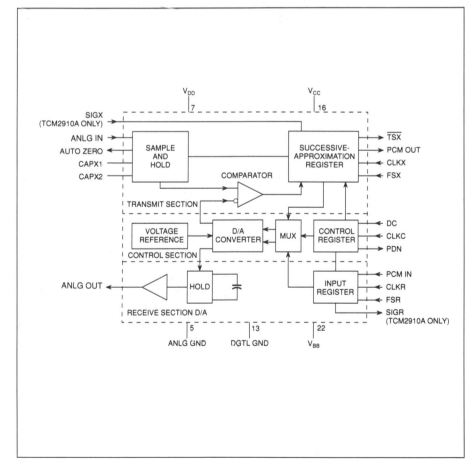

The TCM2910A can be considered as three separate subsections: transmit, control, and receive. A voltage reference in the control section provides a precision voltage reference for D/A circuitry. The D/A converter serves two important functions in the codec. First, it converts and expands digital PCM data into an analog output signal sent to the filter. Second, the D/A circuit is also used as part of the A/D converter which translates filtered analog signals from the subscriber loop into compressed PCM data for transmission. Other control circuitry switches the codec between transmit and receive modes.

Two additional features are built into the TCM2910A. An external power-down control allows the codec to be turned off when not in use. In large systems, this can result in substantial power savings. External time-slot allocation control is also provided. This flexibility enables the microprocessor to direct the timing of transmitted and received words in the codec. *Figure 7-13* shows interconnections between the TCM2910A and TCM2912C.

COMBINED CODEC AND FILTER IC

The "combo" codec combines codec and filtering functions into one integrated circuit.

Figure 7-14 shows the block diagram of a combo. It is called a combo because it combines the codec and filter functions into one IC. The filter functions can be recognized quickly by comparing *Figure 7-14* to *Figure 7-9*, and the related functions of *Figure 7-10* can be identified easily for the codec of *Figure 7-14*.

The TCM2914 is compatible with the TI D-type channel banks and asynchronous clocks used by AT&T. The TCM2913 is designed for use with synchronous clocks. Either μ-law or A-law companding can be selected with the μ-law/A-law select pin. Unless this pin is tied to the V_{BB} supply to select A-law, the companding is by μ-law.

In Chapter 6, *Figure 6-7* showed how signaling occurred in the bit stream. It showed the 8th bit of the code word being used for signaling and that this bit signaling occurred every 6th frame. This is called A signaling. There is also a B signaling per channel giving the possibility of four different conditions being identified by the A and B signaling per channel. B signaling is also the 8th bit, but it is sent in the 12th frame. Thus, signaling is sent every 6th frame and the framing bit is used to identify the 6th, 12th, 18th, 24th frame, etc. Virtually all μ-law codecs provide for the insertion of the A and B signaling bits. On the TCM2914, the insertion for transmit is made on the μ-law/A-law pin when μ-law companding is selected and the signal output for the receive channel is on the Signaling Bit Out pin in *Figure 7-14*.

**Figure 7-13.
Typical Filter/Codec
Interface.** *(Courtesy
Texas Instruments, Inc.)*

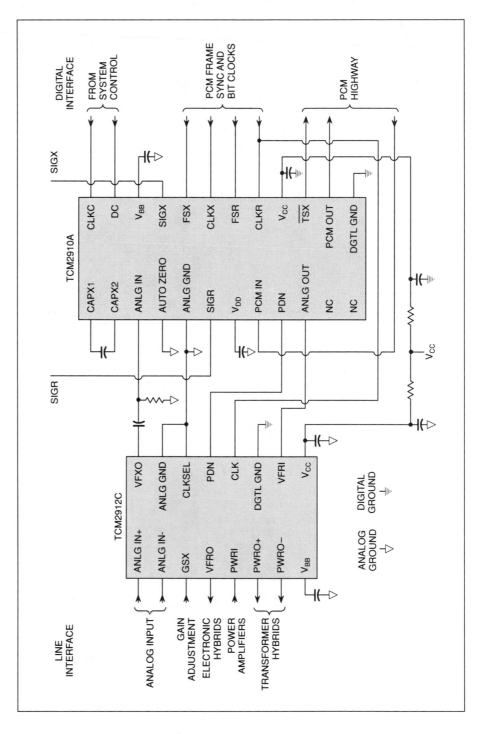

**Figure 7-14.
Combined Single-Chip
PCM Codec and Filter
(TCM2913 and
TCM2914)**

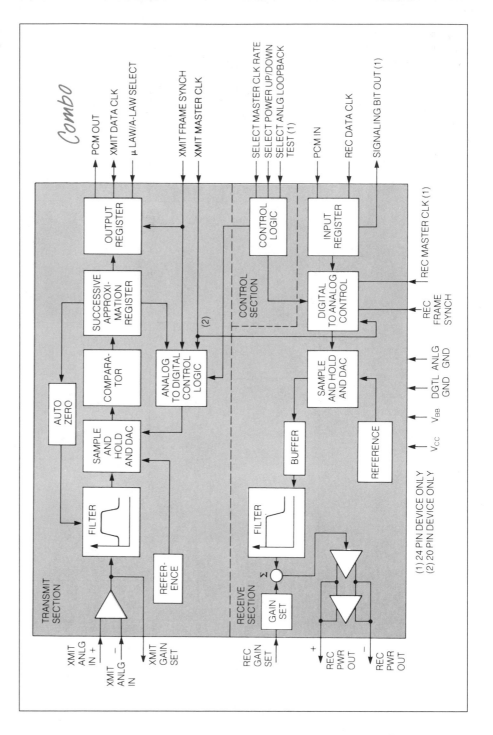

DTMF Receivers

DTMF receivers are used to decode the dual-tone dialing signals sent to the central office. The difficult task of decoding these unique signals once required complex, expensive circuitry. New DTMF decoders can be fabricated onto a single IC.

Dual Tone Multi Frequency tones and tone generator circuits were discussed in Chapter 4. With few exceptions, DTMF generators are easy to understand and design. However, those tones generated in the telephone set must be interpreted at the central office and converted to appropriate switching signals. Although tone decoding itself is not terribly difficult, the special nature and characteristics of DTMF tones present unusual problems for engineers. Until recently, DTMF decoding equipment required complex and expensive electronics which had to be shared with as many as 30 subscriber lines.

As DTMF tones are used more and more for data communications, the demand for DTMF receivers will increase. With the increased manufacturing volume, their cost should decrease.

Figure 7-15a again shows the DTMF frequencies generated at the telephone set. The frequencies in each horizontal row (low-frequency group) and each vertical column (high-frequency group) are separated by intervals of approximately 10%. The low-frequency and high-frequency groups are separated by about 25%. These specific frequencies have been selected with a great deal of care to meet several criteria, but one of the more important requirements is to have the minimum amount of harmonic interaction.

A DTMF receiver must do the following:

1. Detect the tone-pair signal properly if the frequencies are within ±2% of the nominal values, and reject the signal if the frequencies are outside the limits of ±3%.

2. Make sure that one and only one tone is present from each group, and that the tone duration is at least 40 milliseconds (ms).

3. Detect as two separate signals any valid tone pairs which are separated by 35 ms or more. Detect a tone pair signal as the same signal, not as two distinct signals, if the separation between the tone pairs is 5 ms or less.

4. Properly detect tones whose level may vary over a 27.5-dB dynamic range. If the two fundamental frequencies of the DTMF pair have a difference in amplitude, it is called twist. The DTMF circuit must detect the tone pair with up to 6 dB of twist.

5. Properly detect DTMF signals in the presence of speech and noise.

Filtering and Detection

It should be obvious from the preceding requirements that filtering is an important function of the DTMF receiver. *Figure 7-15b* illustrates the frequency response of an ideal DTMF filter. Such a device is called a bandsplit filter because its output is in two separate bands of frequencies to pass the high and low groups of DTMF tones while rejecting other frequencies. The output of the filter must go to a detector to accomplish the other requirements.

**Figure 7-15.
DTMF Frequencies
and Filter**

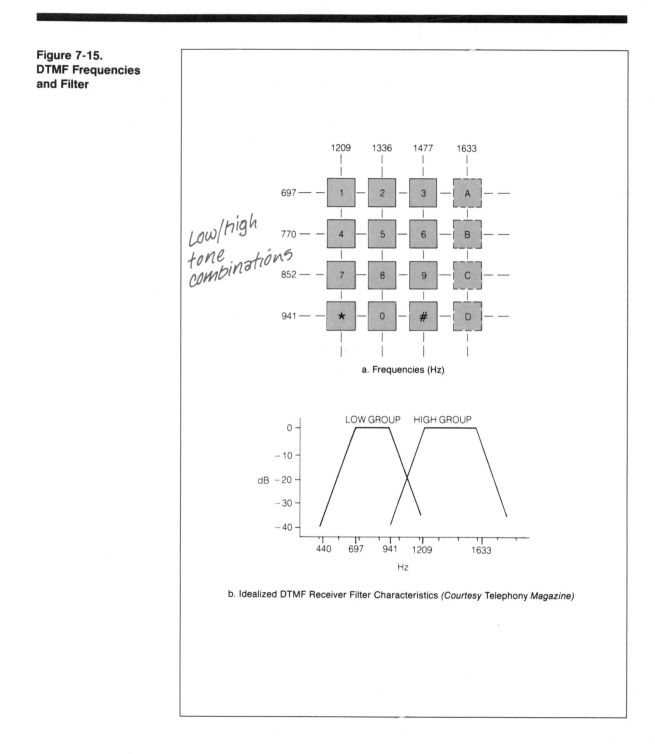

a. Frequencies (Hz)

b. Idealized DTMF Receiver Filter Characteristics *(Courtesy* Telephony *Magazine)*

The filtering section provides for gain, and noise and speech rejection.

Figure 7-16 is a generalized block diagram of a bandsplit filter feeding a detector in order to accomplish the two required functions. Semiconductor manufacturers are producing integrated circuits to provide both the filter and the detector functions on a single chip, such as the Silicon Systems 75T201. The filter functions in the IC are similar to those used in the digital line card in the encode and decode path. It provides signal gain, input filtering for noise and speech rejection, and separate (bandsplit) filtering for the group signals. After filtering, the frequency signals are formed into square waves before passing to the detector.

**Figure 7-16.
DTMF Detection**

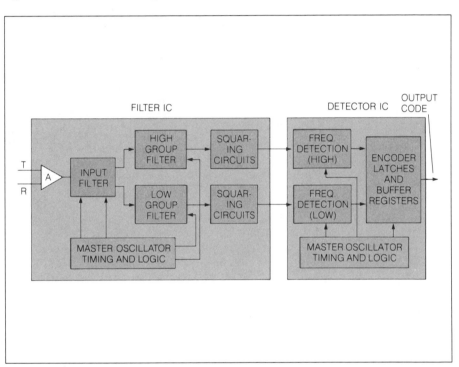

Each DTMF frequency is tested to ensure its validity and if valid is outputted as a code to represent the dialed number.

At the detector, each group signal is processed separately. Digital techniques are used to determine frequency by counting the number of master clock pulses present in each period of the unknown frequency. The detected frequency of the high group and that of the low group are logically tested to determine if they are valid DTMF frequencies. If so, they are combined to produce a coded output representing the dialed number.

The detector must reliably recognize valid DTMF tone pairs from "talk-off" tones without utilizing too much detection time.

One of the more difficult tasks of the detector is to determine if a tone within the DTMF bands is really a DTMF tone or is a sound produced by speech that merely resembles a DTMF tone. If the "accept" criteria are too relaxed, DTMF tone-pairs are quickly recognized as valid, but so are speech segments or other sounds which "sound like" a real DTMF signal. Such detection is referred to as "talk-off," and can lead to wrong numbers or other erroneous operation at the receiver. If the "accept" criteria are too strict, detection time is stretched and the receiver no longer meets timing requirements. The result is a compromise between the effects of noise overriding true signals and the chance of talk-off errors. The detection time is typically about 10 milliseconds.

An Integrated DTMF Receiver

The intricate filtering and timing requirements needed to build accurate, reliable DTMF decoders added a great deal of complexity, size, power consumption, and cost to central office circuitry. However, the same advances in integrated circuit technology which have made single-chip telephones possible have also made single-chip DTMF receivers a reality. Silicon Systems, Incorporated (SSI) manufactures the 75T line of integrated DTMF receivers.

Figure 7-17 shows the complete block diagram for the SSI 75T201 DTMF Receiver. It is designed to interpret signals from 3×4 or 4×4 keypads. Amplification, band-split and bandpass filtering, wave squaring, regulator, oscillator, and decoding/latching logic are all incorporated on the same 22 pin DIP (Dual In-line Package) IC. The only external components required to run the decoder are a 3.58 MHz (color burst) crystal and two power bypass capacitors.

DTMF signals from the subscriber loop are sent to the analog preprocessor input. Signals are amplified and processed through an initial bandpass filter for frequencies from 500 Hz to 6 kHz. Preprocessed signals are then split into two bands—each containing only one tone group. Row and column frequencies are now separate.

Zero crossing detectors create perfect square waves of the major frequency at each band-split filter. When a true DTMF tone is present for a sufficient amount of time, the resulting square wave will have enough amplitude to be processed by a network of bandpass filters. These bandpass filters will discriminate the particular tone frequency. When a valid DTMF signal is present, one output from each bandpass group will be great enough to exceed the amplitude detector reference voltage. Timing circuitry will latch under these conditions and detect the bandpass outputs that are active. Decoding logic produces a 4-bit output in either hexidecimal or binary coded 2-of-8 code as shown in *Table 7-1*. A single logic input can be used to switch between output codes.

A DTMF output decoder circuit is shown in *Figure 7-18*. The circuit can convert a binary coded 2-of-8 output to an actual 2-of-8 output. It will operate with DTMF signals generated from a 3×4 or 4×4 keypad. Signal line DV indicates that a valid decode procedure has taken place, and data will remain available at the output until a valid dialing pause occurs, or until the Clear Line (CLRDV) is made logic high.

**Figure 7-17
SSI 75T201 Block
Diagram.** *(Courtesy
Silicon Systems, Inc.)*

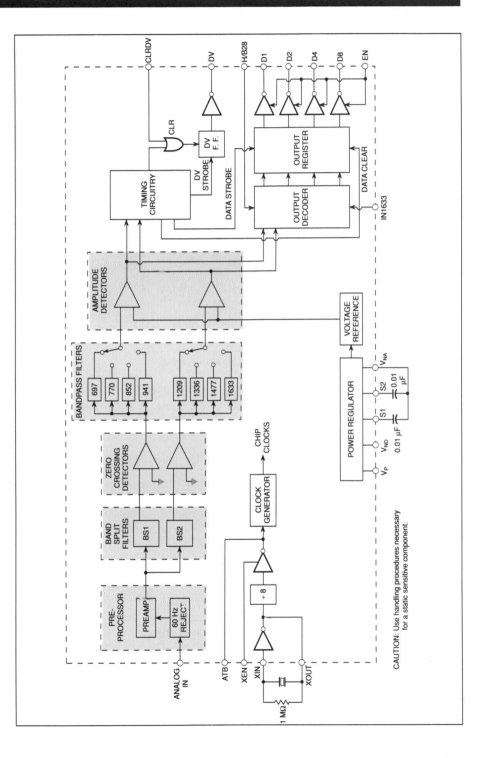

**Table 7-1.
SSI 75T201 Output
Codes***

	Hexadecimal					Binary Coded 2-of-8			
Digit	D8	D4	D2	D1	Digit	D8	D4	D2	D1
1	0	0	0	1	1	0	0	0	0
2	0	0	1	0	2	0	0	0	1
3	0	0	1	1	3	0	0	1	0
4	0	1	0	0	4	0	1	0	0
5	0	1	0	1	5	0	1	0	1
6	0	1	1	0	6	0	1	1	0
7	0	1	1	1	7	1	0	0	0
8	1	0	0	0	8	1	0	0	1
9	1	0	0	1	9	1	0	1	0
0	1	0	1	0	0	1	1	0	1
*	1	0	1	1	*	1	1	0	0
#	1	1	0	0	#	1	1	1	0
A	1	1	0	1	A	0	0	1	1
B	1	1	1	0	B	0	1	1	1
C	1	1	1	1	C	1	0	1	1
D	0	0	0	0	D	1	1	1	1

Courtesy Silicon Systems, Inc.

Single-chip decoders offer a reliable, inexpensive, and efficient replacement for older central office equipment. They are economical enough to use a dedicated decoder for each subscriber loop instead of sharing the decoder among many subscribers.

ELECTRONIC CROSSPOINT SWITCHING

In Chapter 1, the crosspoint switching was shown to be supplied by electromechanical crossbar switches and by reed relay switches. Reed relays are still being used by some telephone companies in their new equipment; however, most manufacturers have adopted some form of *true* electronic switching. An example of such a device is shown in *Figure 7-19.*

The basic switching element is a PNPN semiconductor device that is represented schematically as shown in *Figure 7-19a*. Each PNPN device is equivalent to two transistors, a PNP and an NPN, connected as shown in *Figure 7-19b*. The device has the following characteristics: If a voltage is applied to the device—positive to the anode and negative to the cathode—the device maintains a high resistance and acts like a reverse-biased diode. It remains in this state until the voltage applied between anode and cathode is increased past the breakdown voltage or until a current is drawn from anode to gate. When a current flows from anode to gate, the resistance of the device between anode and cathode becomes very low and the device conducts like a forward-biased diode.

The PNPN semiconductor device is used for electronic crosspoint switching to replace the older traditional mechanical switches. It is normally biased to inhibit conduction, but when the bias is changed to accomplish switching, the device conducts.

**Figure 7-18.
DTMF to 2-of-8 Output
Converter.** *(Courtesy
Silicon Systems, Inc.)*

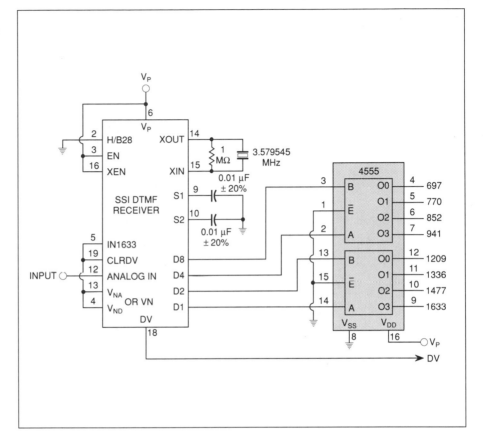

A so-called holding
current maintained by
the loop current
source, keeps the de-
vice in conduction.

Now, even if the current from anode to gate is turned off, the PNPN device maintains its low-resistance state until the voltage from anode to cathode is reduced to the point where a certain minimum current, called the holding current, cannot be maintained through the device. When the current through the device falls below the holding current value, the device changes back to its high-resistance state.

The device used in the No. 10A is manufactured in integrated circuit form[1]. As shown in *Figure 7-19c*, 32 crosspoints are fabricated in an 8-column by 4-row matrix. It is used as a low-resistance connection between the row and column leads by drawing a current out the gate lead while maintaining a minimum holding current from anode to cathode.

[1]*J.M. Adrian, L. Freimanis, R.G. Sparber, "Peripheral Systems Architecture and Circuit Design", The Bell System Technical Journal*, Vol. 61, No. 4, (April 1982), pp. 451ff.

**Figure 7-19.
Crosspoint Elements
and Array**

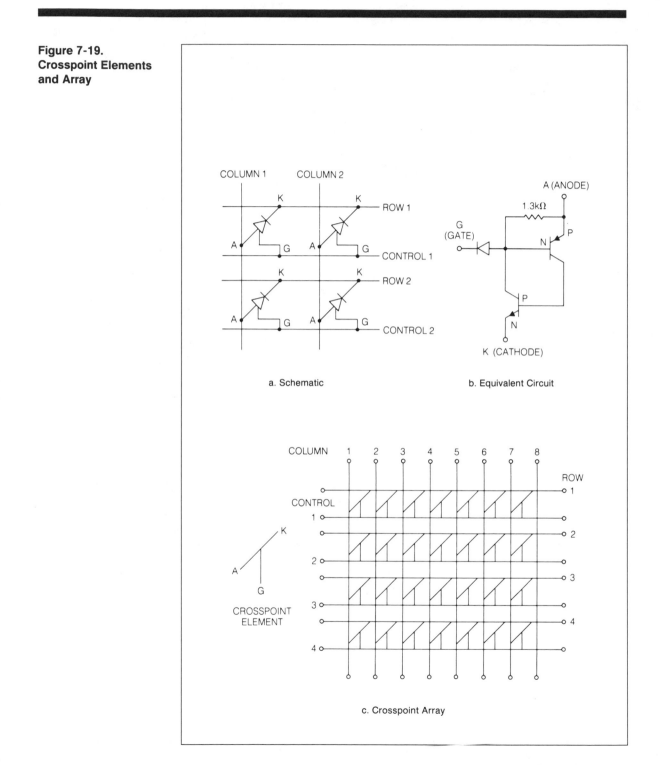

a. Schematic

b. Equivalent Circuit

c. Crosspoint Array

Switching arrays are used to connect two subscriber lines.

Figure 7-20 demonstrates how the arrays are used to connect two-wire balanced speech transmission between two subscriber lines. Only partial arrays are shown and the detailed circuitry for the bottom half of *Figure 7-20* is just like the top half. The circuit connection is completed in two halves. Initially all points in the path are at a high voltage and all current sources I_h and I_t are off. The symbols for the current sources mean they supply a constant current of value I_h or I_t. I_h is the holding current and it is above the minimum value to hold the devices in a low-resistance state. It is the talking current that flows through the subscriber loop.

The connection is made as follows: First, I_h for the top section is enabled. Second, the logic gates, either G_A, G_B, or G_n turn on so that current can be drawn from gate to anode on the respective crosspoint elements. Note that any combination of crosspoints may be used to establish a path through the arrays. Third, the logic gate, G_1, controlling the driver for the center-tap of the input transformer turns on and the path is completed for the I_h current to be conducted to ground. This turns on and latches the PNPN devices, then the crosspoint logic gates are turned off. Now all crosspoints in the enabled path (shown in bold) are at low resistance and the top-half connection is completed. The same steps occur for the bottom-half circuitry of *Figure 7-20* to complete the switching that connects the calling subscriber line to the called subscriber line. Current sources I_t are switched on to support conversation on the lines. Speech is coupled from the top-half to the bottom-half by capacitors, but the dc is blocked.

The path connections are released by turning off the center-tap driver, disabling the current sources, and returning lines to a high voltage.

An Integrated Crosspoint Switch

Silicon Systems manufactures the SSI 78A093, a 96 crosspoint switch configured as a 12×8 array shown in *Figure 7-21*. Seven address lines (3 column address lines—Ay0 to Ay2; 4 row address lines—Ax0 to Ax3) are available under microprocessor control to access each of the 96 switching points. A switch is selected by applying its appropriate row and column address to the switch. Open (off) or close (on) data is entered via a Data pin and latched by a positive pulse on the Strobe line. If Data is logic low, a strobe will turn the addressed station off, and vice versa.

The integrated crosspoint switch offers many advantages over its discrete, relay-type predecessor. First, power consumption is very low (under 100 milliwatts for the entire device) due to its CMOS fabrication. Analog switches provide better than –92 dB of crosstalk immunity, only 28 ohms of on resistance, and introduce almost no distortion to the signal.

**Figure 7-20.
Electronic Crosspoint
Connection**

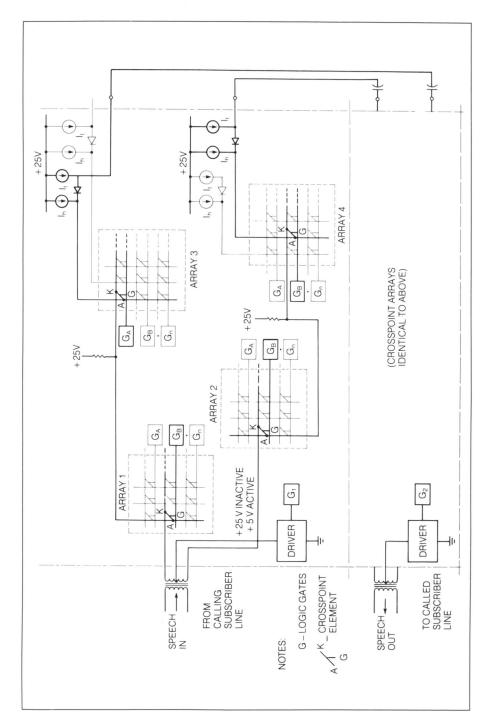

**Figure 7-21.
SSI 78A093 Block
Diagram.** *(Courtesy
Silicon Systems, Inc.)*

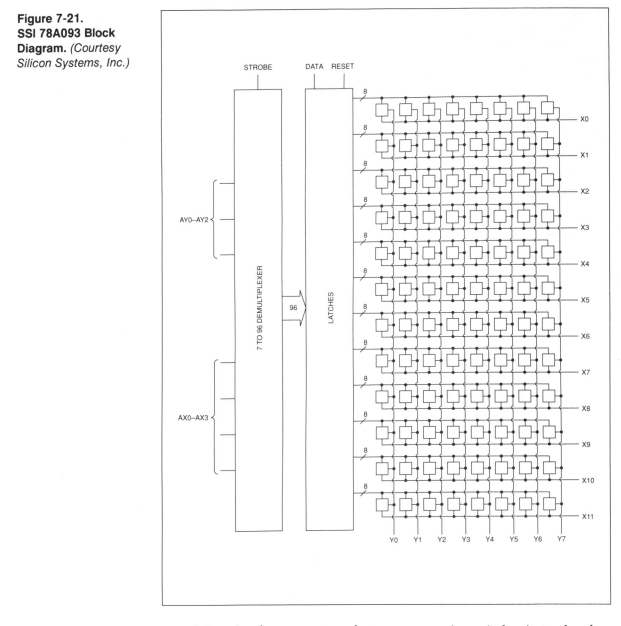

It is a simple matter to apply many crosspoint switches in tandem by adding a selector circuit to manipulate the Strobe signal. *Figure 7-22* shows a basic example of a multiswitch network.

**Figure 7-22.
Crosspoint Switch
Application.** *(Courtesy
Silicon Systems, Inc.)*

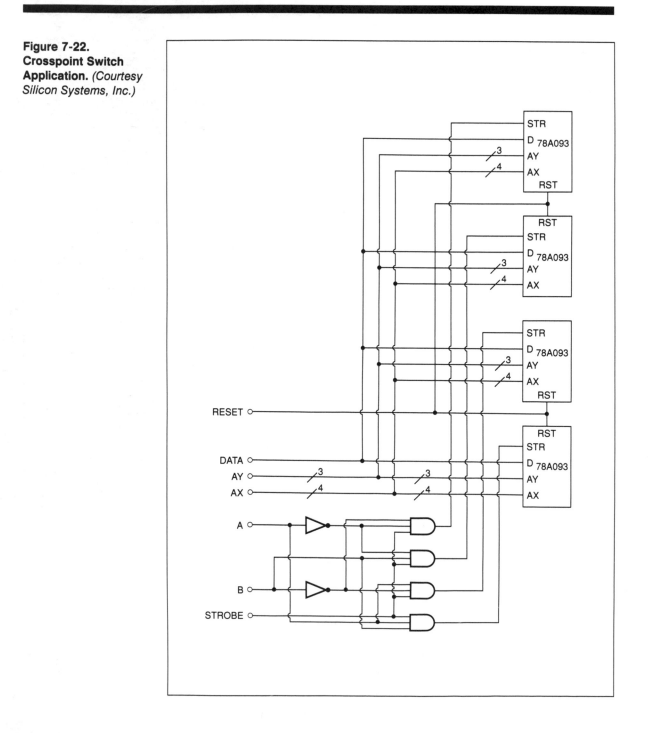

SUBSCRIBER LOOP SYSTEMS

State-of-the-art electronics are being used to greatly extend the operating distances of local loop lines, by using line amplifiers and by restoring current and voltage signaling levels.

One of the main applications of electronics in the subscriber loop is to extend service when the local loop becomes very long. As discussed in earlier chapters, all functions of local loop operation become marginal when loop resistance exceeds 1,300 ohms and loop current drops below 20 mA. In particular, DTMF dialing becomes unreliable, especially at low temperature, and transmitter gain suffers. The problems are most noticeable in the rural areas where line lengths of ten miles and longer occur frequently.

Electronics helped in the past by providing lower cost amplifiers to boost signal levels and by restoring voltage or current levels required for dialing and signaling. Newer techniques include multiplexing several (4 to 8) subscribers onto one local loop as discussed in Chapter 6.

More recently, microprocessors are being used in range-extender systems to provide automatic voice and signal gain at lower per channel cost. These systems automatically adjust themselves to provide the best transmission based on instantaneous measurements made at the time of call setup.

DIGITAL MULTIPLEXING EQUIPMENT

Chapter 6 introduced important concepts of T1 digital transmission using the DS-1 multiplexing format. Early T1 transmission circuitry required as much as 200 discrete integrated circuits to provide every needed function, as well as meet the demanding specifications of U.S. and European networks.

Rockwell International manufactures the R8070 T1/CEPT PCM transceiver which integrates all necessary functions for a pulse code modulation receiver and transmitter onto a single IC chip. The R8070 provides synchronization, monitoring, and signal extraction circuitry. *Figure 7-23* shows a complete block diagram for the R8070.

Eleven modes of operation are available to support most common format variations of T1 and CEPT formats. A microprocessor can interface easily with the R8070 to enhance control and the serial or parallel transfer of data. This device represents a tremendous savings in component costs and system maintenance. Although a full description of a complete T1 application is beyond the scope of this chapter, *Figure 7-24* is a diagram outlining a typical component configuration using R8070 and its associated 4-wire interface device, the 8069.

Digital Subscriber Loops

Digital techniques throughout the system, from the subscriber's telephone set through the switching office, provide the highest quality service at the most cost-effective price.

Digital carrier systems combined with local digital switching have eliminated the need for a multiplexer for the longer local loops and the conversion to an all-digital subscriber loop will aid this even further. In the previous discussions of digital transmission, the digital carrier brought the digital signals to the central office, where they interfaced with an analog subscriber loop. Digital subscriber loops allow the binary digits to come all the way to the subscriber's telephone set because all information

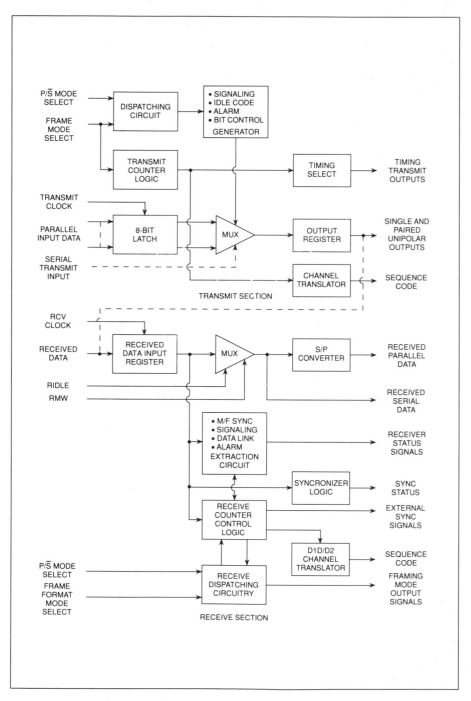

**Figure 7-24.
Typical R8070
Application**

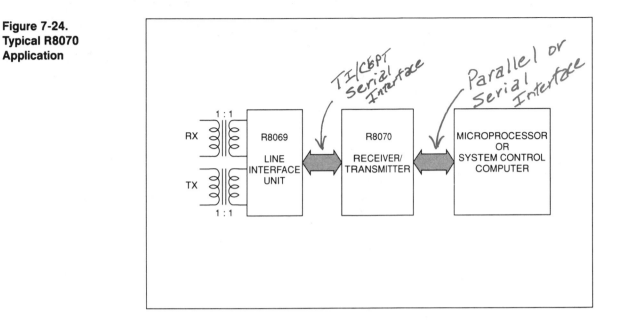

is handled digitally. The telephone set itself contains the codec and filtering functions. Voice, data, supervision, and signaling are all in the same form and will be switched, transmitted, and processed over the same distributed system. Such systems already exist in some PABX systems and eventually will be available for the public network.

WHAT HAVE WE LEARNED?

1. The subscriber line interface requires a wide variety of functions described as BORSCHT functions.
2. BORSCHT describes *B*attery feed, *O*vervoltage Protection, *R*inging, *S*upervision, *C*oding, *H*ybrid, and *T*est functions.
3. Integrated circuits are being designed to accomplish as many of the BORSCHT functions as possible on one chip.
4. Codecs are circuits which are being designed in integrated circuit form that do the digital encoding and decoding on the same chip.
5. Both high-pass and low-pass filters are required to reject noise and unwanted signals and to smooth the D/A output signal to reproduce the speech.
6. Detecting DTMF tones at the receiver is much more difficult than generating the tones at the telephone set because the receiver must be able to respond to real tones, but reject spurious signals that resemble tones.
7. A bandsplit filter has two separate passbands to separate the high-group and low-group frequencies of the DTMF tones.
8. PNPN semiconductor devices are being used to perform the cross-point switching function instead of reed relays. The PNPN devices are being made as integrated circuit arrays.

Quiz for Chapter 7

1. The designer of central office equipment should:
   a. assume electronic telephones in metropolitan exchanges.
   b. change only one function of the interface at a time.
   c. maintain the standard interface between the subscriber and the CO.

2. The difference between the lineside and the trunkside interfaces is that:
   a. trunkside has the requirement of BORSCHT, and lineside does not.
   b. lineside interfaces are usually multiplexed analog carrier systems.
   c. lineside refers to the subscriber loop, and trunkside to connections to other switches.

3. The SLIC is:
   a. the speech loop interface card.
   b. the subscriber line interface circuit.
   c. the subscriber loop input command.

4. BORSCHT, as used in this book, stands for:
   a. battery, overvoltage, ringing, supervision, coding, hybrid, and test.
   b. a Russian beet soup fed to telephone company employees who work outdoors.
   c. battery, on-hook, ringing, subscriber interface, coding, hybrid, and trip.

5. The principal difference between loop start and ground start is:
   a. loop start uses relays; ground start uses integrated circuits.
   b. loop start is one-way signaling which may involve a slight delay; ground start is either-way signaling that is instantaneous.
   c. loop start uses the line and cutoff relay; ground start uses the A, B, and C relays.

6. Foldover distortion or aliasing is:
   a. eliminated by filtering out frequencies below 300 Hz.
   b. another name for crosstalk.
   c. the presence of spurious frequencies caused by having too high frequencies in the sampled signal.

7. Codecs usually encode the signal by:
   a. sampling it and outputting the amplitude of the sample.
   b. folding the signal over and encoding it.
   c. sampling it and using successive binary approximations to encode.
   d. comparing analog waveforms to obtain the closest match.

8. DTMF receivers are:
   a. fortunately much simpler than DTMF generators.
   b. required to tell the difference between random speech frequencies and DTMF tones.
   c. designed to accept the digit if either of the two tones is present, to allow a margin for error in transmission.

9. Semiconductor crosspoint switching arrays:
   a. use PNPN devices that will remain latched in the conducting state.
   b. are used in the Bell No. 1A Electronic Switching System.
   c. are not yet practical, because of the high voltages that must be switched.
   d. are another name for subminiature reed relays.

10. Electronic crosspoint arrays require:
   a. a minimum holding current and a talking current to stay latched.
   b. a minimum holding current to stay latched.
   c. the difference between holding and talking current to operate.
   d. a voltage greater than 48 V between anode and cathode to turn off.

Network Transmission

ABOUT THIS CHAPTER

This chapter describes the techniques, transmission link (media), and equipment used to transmit the digital signals described in Chapters 6 and 7 over long distances. It describes channel banks, multiplexers, and repeaters which are part of the electronic equipment used in digital transmission systems.

WHY DIGITAL TRANSMISSION?

Digital transmission techniques are becoming increasingly cost-effective, even for short distances. Also, the quality of digital methods is superior to analog.

In Chapter 6, a number of advantages and disadvantages of using digital systems were discussed. Despite the disadvantage of requiring eight times the bandwidth, the use of digital techniques is increasing rapidly in the network. Increased performance at lower cost using integrated circuits was pointed out as a prime advantage. This is having such an impact that shorter and shorter digital transmission lines between different types of offices are becoming cost effective. Lines as short as 10 miles are already economical, and if the switching equipment at each end is digital, digital transmission lines of any length will be practical. Less noise, lower signal to noise ratio requirements, and lower error rates are additional advantages.

**Table 8-1.
FDM Versus PCM
Noise Comparison**

	FDM/Radio/Cable	PCM/Radio/Cable
Multiplex	2,500 pWp*	130 pWp equivalent
Radio/cable	7,500 pWp	0 pWp
Total	10,000 pWp	130 pWp equivalent

*pWp = peak picowatts (10^{-12} watt) weighted according to the psophometric weighting curve
(Source: R.L. Freeman, Telecommunications Transmission Handbook, Second Edition, John Wiley & Sons, 1981, Copyright © 1981 by John Wiley & Sons, Inc., Reprinted by permission.)

Data from Freeman, listed in *Table 8-1,* indicates that the total noise generated in a 2,500 kilometer (km) standard frequency division multiplexed (FDM) system by the transmission media (cable and radio) is 10,000 pWp. The noise in an equivalent 2,500 km PCM transmission system is only 130 pWp—a reduction of almost 19 dB. By adding repeaters in the line, the error rate is reduced. If the error rate is too high, spacing repeaters closer together will lower it. Thus, transmission errors become independent of total transmission line length.

The digital transmission line is more versatile and is favored by engineers.

A few additional advantages are worth noting:

1. Digital techniques contribute to the design of rugged transmission lines that are easy to design, easy to install, and easy to maintain.
2. Digital transmission lines handle all varieties of source signals—broadcast music, data, and television—which will eliminate the need for special lines for special signals.

DIGITAL CHANNEL BANKS

Digital channel banks combine many voice inputs into one wideband digital channel.

Recall that in Chapter 6, channel banks were defined as the equipment at each end of a digital transmission that converts speech signals into coded bits. Equipment cost analysis has dictated the development of equipment that multiplexes many channels onto one wide bandwidth channel. For example, the standard system (T1 carrier) is made up of 24 individual voice channels. The channel banks generate and multiplex the digital signals. The Bell System channel banks are called D-type (for digital) and form the foundation for a whole series of multiplexers.

D-Type Channel Bank

The D-type channel bank generates an output, placing the digitized voice signals into their assigned time slots.

The output from a D-type channel bank is a 1.544 Mbps digital signal called a DS-1 signal. The DS-1 signal might be transmitted on a T1 transmission line, which has repeaters in the line for continued signal regeneration, or it may be multiplexed with other DS-1 signals to produce a still higher level multiplexed signal. The DS-1 signal contains the bit stream with the digitized voice signals from each channel in their time slots and the channel signaling bits in positions as discussed in Chapter 6 (*Figure 6-6*)

D1 Channel Bank

The D1 channel bank performs both transmitting and receiving. Its functions are encoding and decoding, and adding and extracting signaling and framing information. It interfaces with the line and therefore multiplexes and demultiplexes the signal.

Figure 8-1 is a block diagram of a section of a D1 channel bank. *Figure 8-1a* shows the transmitting portion, and *Figure 8-1b* shows the receiving portion. The transmitting portion performs three basic functions:

1. Digitally encoding the 24 analog channels
2. Adding the signaling information from each channel
3. Multiplexing the digital stream onto the transmission medium.

The channel inputs, gates, filters, compression circuits, and encoder provide filtering, sampling, compression (half of the companding function), and coding (analog to digital conversion). The signaling information from the channel signaling circuits and the framing pulses from the master clock and framing generator are inserted in the digital stream by the transmit converter. The resulting PCM stream is coupled to the transmission lines through a common control circuit. Since many different signaling schemes as well as both two-wire and four-wire trunks are in use, each of the five kinds of D channel banks [D1, D2, D3, D4, and DCT (Digital Carrier Trunk)] contains interface circuits between individual voice channels and the common equipment shared by all circuits.

**Figure 8-1.
D1 Channel Bank (24
Channels)**

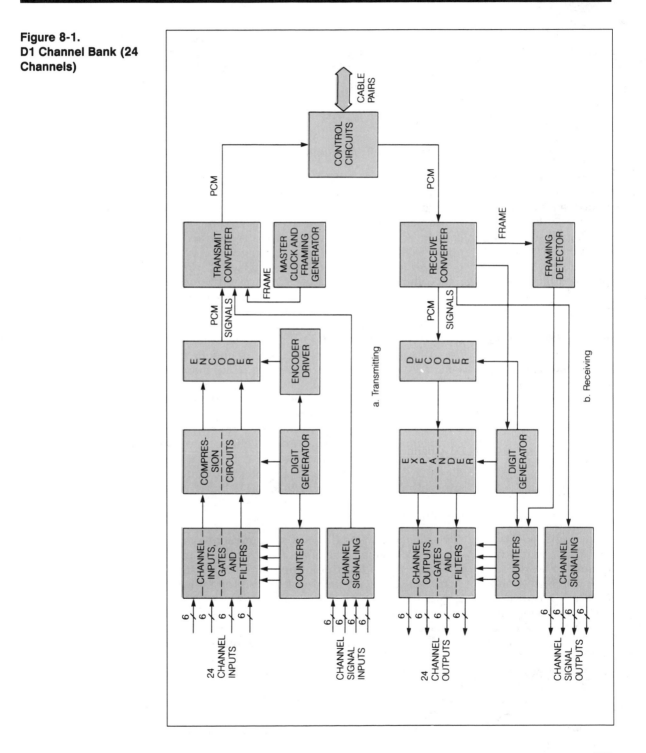

The receiving portion reverses the process. It extracts the signaling and framing information, passes the signaling to the receive side of the channel signaling circuits, and passes the PCM encoded voice to a decoder (which does the digital to analog conversion) and expander (the other half of the companding function), and on to the individual channel filters and output circuits. A D1 channel bank contains three 24-channel sections (72 channels).

D Channel Bank Modifications

The D1 and D2 series channel banks represented improvements over the D1 banks. Packing density was increased (i.e., more channels) and improved channel noise figures resulted. The technique involved a more efficient use of the 8 bits per time slot to carry digitized voice, and the decreased use of unnecessary signaling information.

The success of the D1 banks, first installed in 1962, led to modifications (D1B and D1C versions), and then to a major revision. The original D1A, B, and C banks used 7 bits for each voice sample and one bit in each code word for carrying the signaling. When it became desirable to connect several T1 transmission spans together in tandem, the quantizing noise produced by the 7-bit encoding was too high for satisfactory performance. In addition, it was realized that providing signaling information in every code word is wasteful since 8,000 bits per second was not required to provide the signaling information for a channel because the signaling information did not change that rapidly.

As a result of these conditions, another modification to the D1 series (the D1D) and the new D2 channel bank were developed. The D2 bank uses all eight bits of every time slot to encode the analog signal except for selected frames. Supervisory and signaling information is sent by using the least significant bit from the code word in each channel every sixth frame. This was described briefly in Chapter 6, but is given in more detail in *Figure 8-2*. This produces an effective average code word length over all frames of $7\frac{5}{6}$ bits (8 bits in the first 5 frames plus 7 bits in the sixth). The companding characteristic also was changed to give better idle channel noise performance and to make the code words easier to linearize digitally. The D2 bank increased the packing density to 96 channels in the same space as the 72 channels for a D1 bank.

D3 and D4 Channel Banks

Further advances in digital microcircuitry resulted in the development of the D3 and D4 banks, greatly increasing channel capacity. Next followed the Digital Carrier Trunk (DCT). The DCT system is microprocessor-based, and has far fewer discrete subsystems.

The D3 and D4 banks were motivated by advances in integrated circuits, allowing packaging of 144 channels in a single bay. Following the D4 bank, the application of programmable distributed computing and signal processing to the problem resulted in the development of the Digital Carrier Trunk unit, or DCT. It was developed by the Bell System to be smaller, lower cost, and easier to maintain than the D4 channel bank. As shown in *Figure 8-3*, it differs from the D4 unit in that it has no separate trunk circuits, no separate signal distributor and scanner, and it uses a single kind of channel unit—as opposed to the over 35 unique types available for the D4. It includes programmable microprocessors called peripheral unit controllers that relieve the main switching machine (usually a No. 1 or No. 1A ESS) of the task of sorting out the signaling information on the incoming trunks. The main switch also uses the peripheral unit controller to send the signaling, alarm, and trunk status

**Figure 8-2.
Sixth Frame Signaling
of D1D Channel Bank**

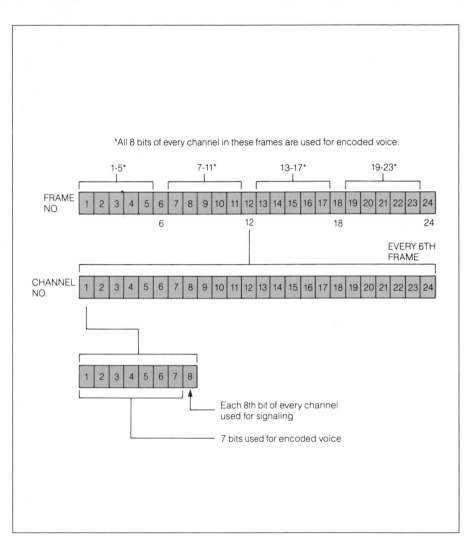

over a data link to a central control unit which routes the voice signals
over the correct trunks.

MULTIPLEXERS

Digital multiplexing is
the technique of com-
bining many digital in-
puts into an output
digital stream.

The electronic circuit that combines digital signals from several
sources into a single stream is called a multiplexer. The basic was shown
in *Figure 6-25*. A multiplexer takes in many channels and allows each in
turn access to the output to produce an output bit stream in serial
sequence. Each input in turn is gated into the proper time slot in the
output. If the time slot for a channel in the output is long enough to
contain the entire input code word, the stream is said to be word
interleaved. If the input is gated to the output one bit from each channel

**Figure 8-3.
Comparison of D-type
Channel Bank and
Digital Carrier Trunk**
*(Courtesy of Bell
Laboratories)*

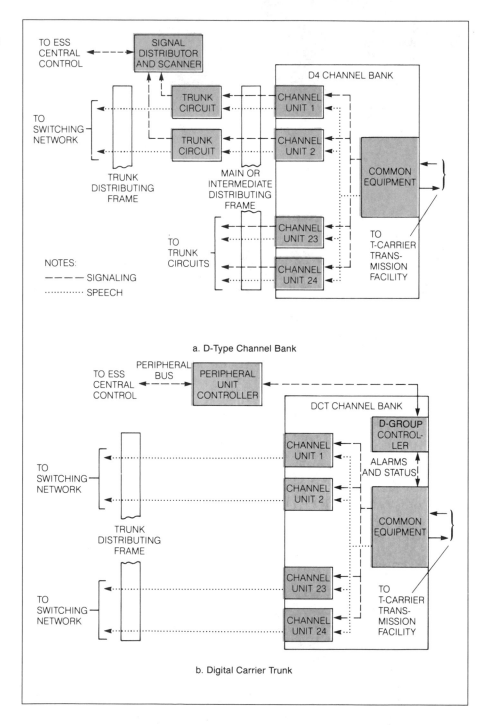

a. D-Type Channel Bank

b. Digital Carrier Trunk

in sequence, the stream is bit interleaved. Since all timing is derived from the master clock, and all of the input streams run at the same rate, the unit is a synchronous multiplexer. All of the higher-order multiplex systems used in the telephone network are the synchronous type, but asynchronous multiplexers are used in data transmission applications.

Framing Formats

In multiplexing, the channels are bunched together in frames and are set off from previous and subsequent frames by a framing bit or framing word.

Message channels are multiplexed into the bit stream in frame formats. Frames are set by the number of multiplexed channels that are packaged together. A D-channel bank multiplexes 24 channels to form a DS-1 signal. CCITT multiplexes 30 channels into a frame. A special code, either a bit, word, or words, is added in the frame to synchronize the frames of the PCM signal. Certain formats have been established for the time placement of the framing information. *Figure 8-4* shows three types—bit framing, bunched word framing, and distributed word framing.

Multiplexer Hierarchy

There is a functional hierarchy based upon the bit rate and channel availability. Lower bit rate and channel availability multiplex units are fed into systems with higher bit rate and channel capacity.

The multiplexers used in the public telephone network are arranged in an order that depends on the bit rate of the PCM signals that are transmitted. Multiplexer units with lower bit rates and fewer numbers of channels couple to units that have higher bit rates and greater number of channels. *Figure 8-5* illustrates the relationship of these systems. The output rates, and, therefore, the numbers of channels that each multiplexer can combine, are matched to the data rates which can be carried on the various transmission media, from the DS-1 signal at 1.544 Mbps to the DS-4 at 274.176 Mbps.

The multiplexing group members in the North American network above the D-type channel banks are:

1. The M1C which combines two DS-1 signals into a single DS-1C signal at 3.152 Mbps.
2. The M12 which combines four DS-1 signals into a single DS-2 signal at 6.312 Mbps.
3. The M13 which combines 28 DS-1 signals into a single DS-3 signal at 44.736 Mbps.
4. The M34 which combines six DS-3 signals into a single DS-4 signal at 274.176 Mbps.

Included in *Figure 8-5* are the different types of transmission lines used from T1 cable to coaxial, waveguide, radio, and optical fiber transmission media.

Figure 8-4.
Framing Formats

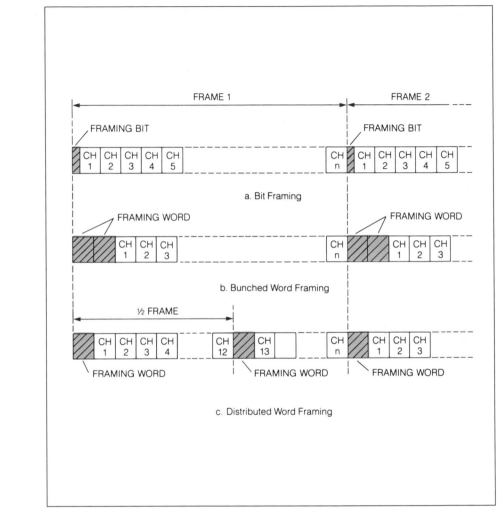

a. Bit Framing

b. Bunched Word Framing

c. Distributed Word Framing

Pulse Stuffing

Variations in the data rates between bit streams inputted to a multiplexer may be offset by pulse stuffing. This technique compensates for frequency differences due to mode clock differences and line propagation delay.

The timing for all operations within a multiplexer, including the clocking for the output bit stream, is provided by a master clock within the unit. The timing for the received data streams, however, is provided by each individual bit stream. The data rate for each of these input bit streams is likely to be slightly different from any of the other bit streams due to variations in the master clocks of the sources transmitting the signals and because of variations in the propagation delays of the circuits and media carrying the bit streams. Some method of allowing for these mismatches must be provided. In multiplexing, the method is called pulse stuffing.

Figure 8-5.
TDM-PCM Hierarchy
(Courtesy of Bell Laboratories)

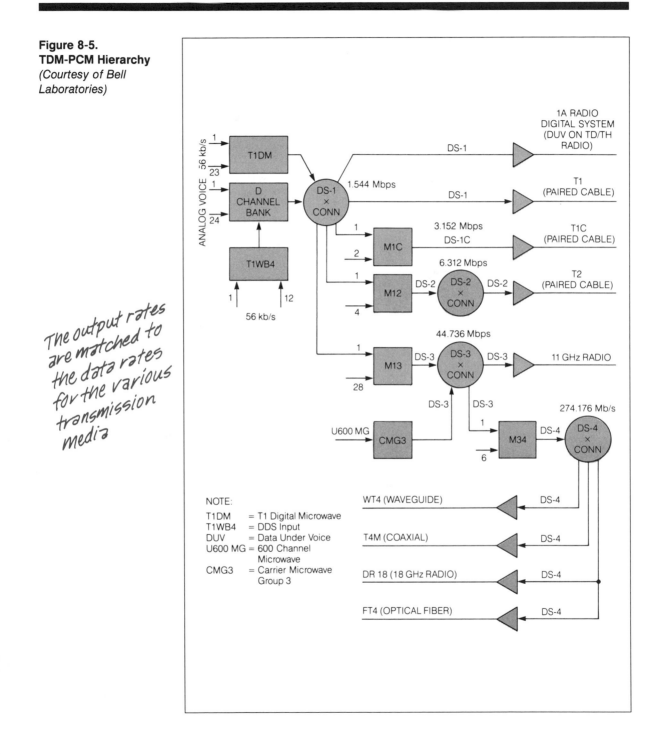

The output rates are matched to the data rates for the various transmission media

NOTE:
T1DM = T1 Digital Microwave
T1WB4 = DDS Input
DUV = Data Under Voice
U600 MG = 600 Channel Microwave
CMG3 = Carrier Microwave Group 3

Figure 8-6.
Pulse-Stuffing
Synchronization
(Courtesy of Bell
Laboratories)

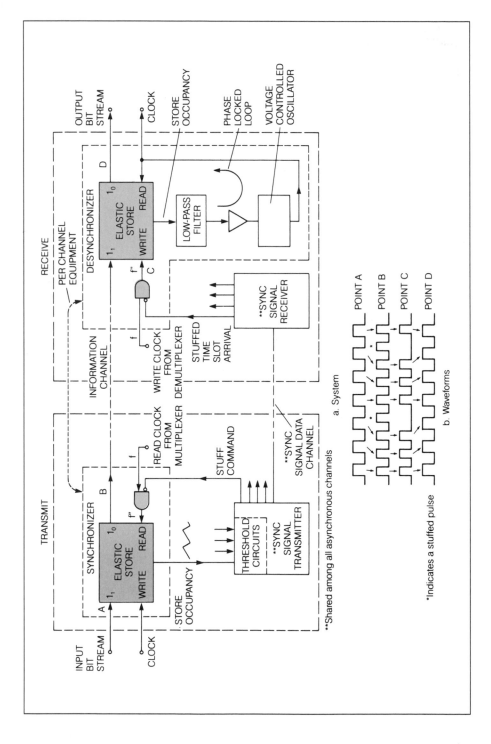

The principle of pulse stuffing, illustrated in *Figure 8-6*, is to cause the outgoing bit rate of the multiplexer to be higher than the sum of all of the incoming bit rates plus any data added in the multiplexer for framing and synchronization. The incoming signals are stuffed with a number of pulses sufficient to raise their rates to be the same as the locally generated master clock. Extra bits (sync signal data channel), common to all channels, are provided to send the receiver the location of the stuffed pulses. The information about the location of the stuffed pulses is sent repeatedly (redundantly) to reduce the probability of an error that would cause loss of frame synchronization.

Each incoming bit stream is fed into a buffer (called an elastic store) at the incoming bit rate (Point A of *Figure 8-6b*) and temporarily stays in a stacked location. At appropriate times, a pulse is stuffed in the stream to provide the output shown at point B in *Figure 8-6b*. Point B's output is transmitted to the receiving unit and stored in another elastic store. A phase locked loop generates the clock which reads out the bit stream at point D. The time slots where pulses were stuffed are identified at the receiver by the stuff codes from the sync signal data channel and the pulses are unstuffed as shown in the waveform of *Figure 8-6b* for point C. The read clock rate of the receiver restores the bit stream to the uniform signal similar to that which was input at point A. If the buffer is ever empty, a pulse is provided so that the output can continue at a constant rate. A corresponding stuff code is sent on the sync channel to identify the stuffed timing bit. Certain formats and rules for detecting stuffing bits for each multiplexer type have been established. Let's look at the ones for the multiplexer for the DS-1C signals.

DS-1C Multiplexed Signal

The master frame of the DS-1C PCM signal is made up of 1,272 bits divided into four 318-bit subframes as shown in *Figure 8-7*. M is the leading bit for the subframes and its pattern of 01 1X identifies the master frame and the subframe boundaries. The X bit is always a 1 unless an error occurs in transmission of the data, in which case the X bit is a 0. The regular bit pattern of 01010101 for F_0 and F_1 establishes the frames with the stuffing bits included for each subframe. The bits designated as stuffing bits are stuffed (interpreted as a null) when the bits shown as C bits are all 1 in a subframe. Otherwise the stuffing bits are interpreted as data. It should be pointed out that the frame lengths for the DS-1C and higher signals are not related in any systematic way with the input DS-1 frames. The multiplexer treats the input DS-1 stream as simply a string of bits, passing data and synchronization pulses alike into the higher level stream.

The higher level multiplexed signals have similar bit patterns established for master frames, subframes, frames, and stuffing bit positions and stuffing rules. As with the DS-1C signal, the bits that form the pattern are distributed across the frame in a uniform way to minimize the probability of missing more than one stuff code due to a burst of errors.

The master frame DS-1C PCM is divided into four subframes, each separated by a pattern identifier code. Stuffing bits have been added to each subframe by following a set of rules set up before- hand.

Figure 8-7.
DS-1C Frame Format
Showing Pulse
Stuffing Bits

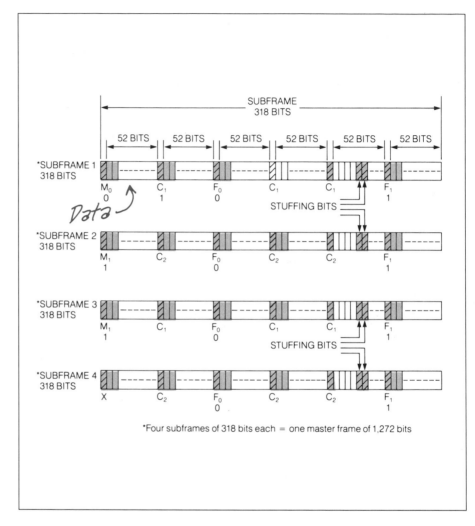

Such a miss would cause a frame slip which would cause a loss of synchronization.

Synchronization

High-speed phase-modulated data transmission is quite vulnerable to synchronization problems.

The previous discussions have emphasized the importance of preventing loss of synchronization, or clock slips. The effect of these errors on digitized voice is not serious until the clock misalignments reach very large values (on the order of 1 part in 100,000). However, digitized signals from voiceband modems and voice or data signals which have been coded to maintain security (encrypted) are highly susceptible to clock slips. High-speed modem data are generally phase modulated, and a single 8-bit slip in the digital data produces a phase shift of 81 degrees. Not only are the data coded for security, but also the data are scrambled further because the modem loses synchronization and may take several seconds to recover. Such slips give audible blips in the voice signal and cause problems in the decoding equipment.

Plesiochronous synchronization utilizes very accurate, but expensive clocks throughout the network, so that timing differences are small.

There are several possible methods for synchronizing a network of independent switching machines such as the telephone network. One method is to supply each machine with a clock so accurate that the difference between it and all other clocks is so small that slips seldom occur. This technique is called *plesiochronous* synchronization. It is expensive to implement since the clocks are expensive and must be redundant at each switch. Nevertheless, this method is the one chosen by the CCITT for international digital transmission in Europe and the gateways to the East. The clocks must be stable to within 1 part in 100,000,000,000.

Another method is to do pulse stuffing network-wide, as is done in the higher-level multiplex systems. This would require every channel at every digital switch to be stuffed separately on transmit. Such an approach is not economically feasible.

In master/slave synchronization, the entire network is locked up to a master clock.

The method adopted for the North American network is that of master/slave synchronization. The master timing frequency reference is maintained in Hillsboro, Missouri, which is near the geographical center of the network. From there, the master timing reference is sent to selected switching centers, which become slave timing centers, over dedicated transmission facilities. From these slave centers, the timing reference is forwarded to lower level centers over existing digital facilities. *Figure 8-8* shows a network diagram for a master/slave synchronizing system for North America with the central reference designated as M. The numbers in the circles are the office classes of the network.

LINE CODING

The line coding format used for sending 1 and 0 states down the line should facilitate the reliable, efficient propagation of information.

The waveform pattern of voltage or current used to represent the 1s and 0s of a digital signal on a transmission link is called the line code. There are many different types of line codes in use and several will be examined in a moment, but first some basic requirements of a line code.

1. The level of direct current in the transmission medium must be constant, and preferably zero.
2. The energy spectrum of the wave must be shaped to avoid frequencies outside of the bandwidth available.
3. The signal must contain adequate timing information so that the clocking signal can be recovered by regenerative repeaters in the transmission path.
4. Errors should be easy to detect, and their detection used to indicate the performance (error rate) of the link.
5. Errors should not propagate; i.e., single-bit errors on the line should not cause multiple-bit errors in the decoder.

**Figure 8-8.
Master/Slave Timing
Network** *(Source:
Luetchford, et al,
"Synchronization of a
Digital Network", IEEE
Transactions on
Communications, Vol.
COM-28, Aug. 1980,
Copyright © 1980 IEEE)*

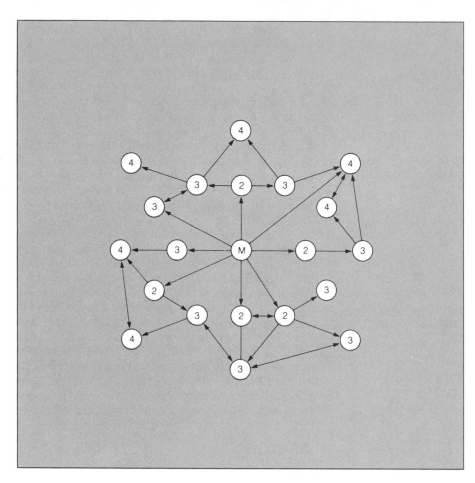

6. The coding scheme should use the code space efficiently; i.e., if possible, the scheme should use only two states to represent one and zero.

7. The coding scheme should minimize crosstalk between channels.

Types of Line Codes

Figure 8-9 details the different types of line codes. In reference to the bit stream shown in *Figure 8-9a*, a typical binary signal waveform is shown in *Figure 8-9b*. It is the simplest of the code types because it uses just two voltage levels; in this example +3 V and 0 V are used. It is called unipolar because it is not symmetrical about 0 V. When the waveform is symmetrical about 0 V as shown in *Figure 8-9c*, it is called a polar code. If the signal level representing each bit (e.g., +1.5 V for a 1 and −1.5 V for a 0) maintains the assigned value for the entire time duration allotted to the bit (time segment T), the code also is called a Non-Return-to-Zero (NRZ) code.

The most common types of line codes are unipolar, polar, and bipolar. The unipolar states are zero and positive, the polar states are positive and negative, and the bipolar states are positive and negative being equivalent states, with zero being the other state.

Figure 8-9.
Types of Line Coding

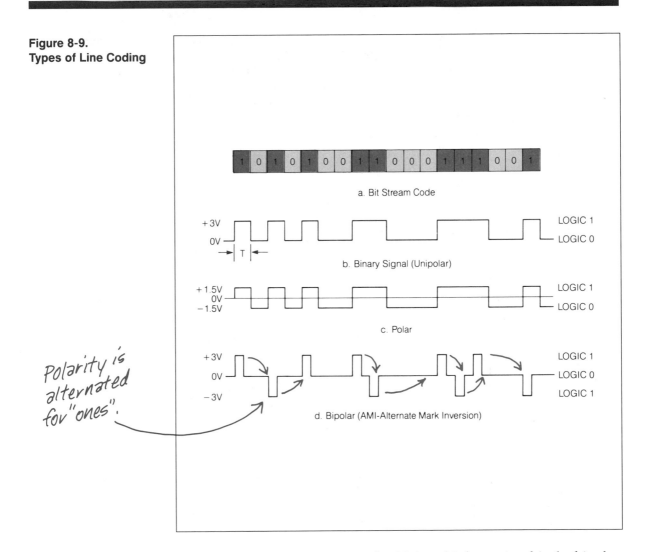

a. Bit Stream Code

b. Binary Signal (Unipolar)

c. Polar

Polarity is alternated for "ones".

d. Bipolar (AMI-Alternate Mark Inversion)

The code that is primary to the DS-1 multiplexer signal is the bipolar code shown in *Figure 8-9d*. It also is called Alternate Mark Inversion (AMI) coding. It produces alternate positive and negative level pulses, symmetrical around 0 V, when successive 1s occur in sequence. As a result, AMI is a three-level or ternary signal, where a 1 is represented by either a positive-going or a negative-going pulse in a signal interval, while a 0 is represented by the absence of a pulse in a signal interval. This scheme, first used by the Bell System in the U.S., satisfies most of the basic requirements listed previously. There is no dc component in the transmitted signal, the amount of energy in the signal at low frequencies is small and, compared with unipolar signaling, AM1 has a substantial advantage in that it has much more immunity to crosstalk (on the order of 23 dB).

The waveform of AMI pulses is shaped to occupy only about one-half of the allowable pulse width. This is done to simplify the circuits that recover the timing information in the regenerative repeaters. The total waveform does not have a 50% duty cycle as occurs in the clock waveform shown in *Figure 8-10a*, but the symmetrical pulses are said to have a 50% duty cycle.

**Figure 8-10.
DC Wander**

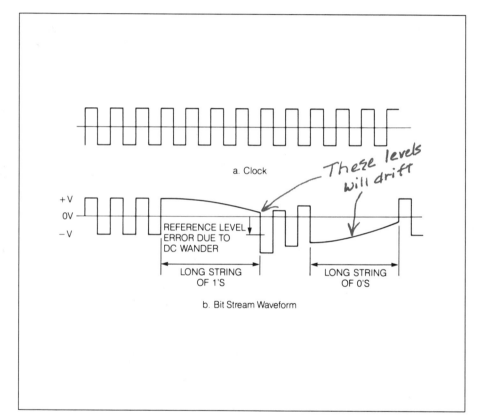

a. Clock

These levels will drift

+V
0V
−V

REFERENCE LEVEL
ERROR DUE TO
DC WANDER

LONG STRING
OF 1'S

LONG STRING
OF 0'S

b. Bit Stream Waveform

DC Wander

Long sequences of ones and zeros over a digital transmission link can cause the dc level to wander. Dc restoration circuits clamp the transmission line to a continuous reference level during these sequences.

A principal requirement of a line code is that it contain no dc, or if a dc component is present, it must be a constant level. This is because most carrier transmission systems do not pass dc. The problem that occurs when a large dc component is present is shown in *Figure 8-10*. Long sequences of 1s or 0s can cause a shift in the waveform amplitude with respect to zero volts. When this occurs, signal recovery circuits that are detecting the 1s and 0s lose proper amplitude reference and errors occur. Special circuits called dc restoration circuits are designed to eliminate the dc wander and restore the proper reference levels.

PULSE AND TIMING RESTORATION

Digital signals are attenuated by an amount which is a function of their frequency.

Digital signals have the useful property of always starting out with a well-defined and unique shape or waveform. The encoding scheme for transmission of the bits down a channel is designed so that only a small number (typically two or three) of such waveforms are possible. This makes it easy to distinguish genuine binary information from noise, even if the waveform is severely distorted. Signals sent down a cable are diminished in amplitude (attenuated) by an amount which depends in part on the frequency of the signal (roughly by an amount equal to 8.6 times the square root of the frequency). If the original signal has a fairly wide bandwidth, the attenuation will be greater for the higher frequencies than for the low.

Distributed line characteristics cause a loss in absolute amplitude, deterioration of waveshape, and phase shifting. These signal deteriorations are mostly corrected for by equalization circuits which compensate for the line parameters and regenerate the signals.

Shifts in the time relationship of signals (phase) also occur as the signals are transmitted over a transmission link. Shifts in phase also contribute to distorting the signal waveforms. The process of compensating for the amplitude and phase shifts is called *equalization*. In principle, the equalization compensation should have an amplitude and phase dependence on frequency which is the exact opposite to that which occurs as the signals are transmitted over the link. However, in practice, the equalization is quite good on amplitude but is not very good at phase correction.

Whenever the original waveforms are reconstructed, the signal is said to be regenerated. The process is not at all like mere amplification, although voltage amplitudes may in fact be increased. The significant fact is that the regenerated new pulses are just like the original ones in size and shape. They may be delayed in absolute time, but have the same position in time relative to each other.

Repeaters

The regenerative repeater converts a weak, distorted digital signal input into an exact replica of the original signal by pulse reshaping and retiming. The signal is actually regenerated, rather than just being "cleaned up."

The system components that perform the regeneration task are called regenerative repeaters. Effectively, they take an attenuated and distorted input signal and reconstruct it at the output to look like the original signal applied to the transmission link. *Figure 8-11* shows the elements of such a regenerative repeater. The two functions it performs are pulse reshaping and retiming. Reshaping restores the pulse shapes to their original form, and retiming sends them out at the proper time intervals. The repeater consists of impedance matching networks, amplifiers, threshold bias circuits, timing circuits, pulse regenerators, and a power supply. Power is supplied down the same pairs as carry the data signals, and is usually either −48-V dc or −130 volts from the central office.

**Figure 8-11.
PCM Regenerative
Repeater**

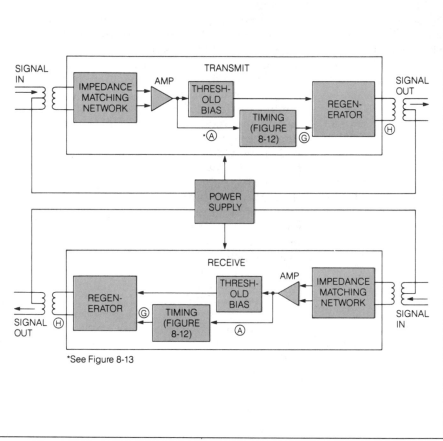

The incoming signal is amplified, and drives the detector threshold bias circuit, timing chain, and balanced pulse regenerator circuits. The threshold bias circuit determines the decision level for the incoming signal, which determines for each pulse interval whether or not a pulse will be produced.

Clock Recovery

In the timing circuit, clock recovery is accomplished by processing the incoming data signal. The inputted signal is rectified, differentiated, and clipped.

The timing chain is illustrated in *Figure 8-12*. Its main purpose is to retime the PCM signal by recovering the clock based on the timing of the incoming signal. The equalized pulse train (balanced about plus and minus levels) is rectified, differentiated, clipped, and input to the resonant tuned circuit. The waveforms of the signals at the points A through G of *Figure 8-12* are shown in *Figure 8-13*. Point A is the input AMI signal, point B is the rectified signal, point C is the differentiated signal, and point D is the clipped signal. Only the positive-going pulses remain to be used as trigger pulses to the tuned circuit.

**Figure 8-12.
Timing Circuit**

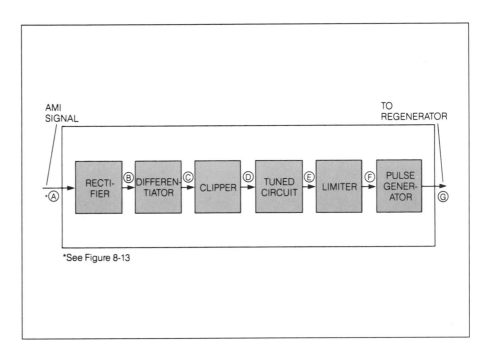

AMI
SIGNAL

TO
REGENERATOR

·Ⓐ RECTI-FIER Ⓑ DIFFEREN-TIATOR Ⓒ CLIPPER Ⓓ TUNED CIRCUIT Ⓔ LIMITER Ⓕ PULSE GENER-ATOR Ⓖ

*See Figure 8-13

The input trigger at point D excites the tuned circuit and causes it to ring (oscillate) at its resonant frequency as shown in waveform E of *Figure 8-13*. The ringing is at a constant frequency, but slowly decreasing amplitude. The constant frequency pulses are counted, divided, and sent through a limiter to produce the clock pulses of waveform F. Each time a trigger pulse arrives at the tuned circuit it restores the amplitude of the ringing signal. It is quite obvious that if a trigger pulse did not arrive for a long period of time, the amplitude of the ringing signal would decrease to zero. This would be the case if a long train of 0s occurred in the bit stream. The ringing amplitude would decrease below the detector threshold and the clock would be lost until the next 1 pulse arrived at the input.

Regeneration

In regeneration, the square waves are inputted to a pulse generator which is used to gate the incoming equalized pulses to the regenerator.

The clock pulses of point F are input to a pulse generator which produces positive and negative pulses at the zero crossing points of the clock square wave. In telephone terms, such clock recovery is termed forward acting and the repeater is said to be self timed. The clock pulses at the positive-going zero crossings of the timing wave are used to gate the incoming equalized pulses to the generator. The pulses from the negative-going crossings turn off the regenerator to control the width of the regenerated pulses. This action is called complete timing with pulse width control. The output at point H in *Figure 8-11* is a regenerated signal as shown in *Figure 8-13* that matches the original generated AMI signal.

**Figure 8-13.
Timing Circuit
Waveforms**

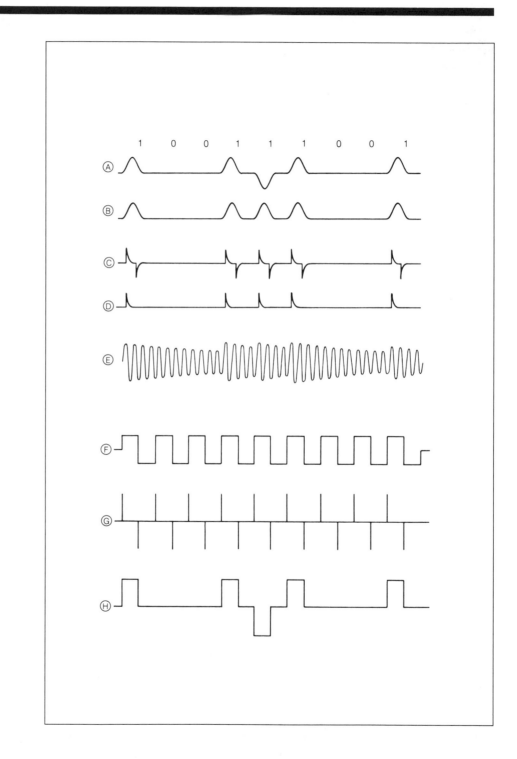

As has been mentioned previously, it is possible to reduce the error rate on digital transmission facilities to any desired value by spacing the regenerative repeaters closer together. As a practical matter, however, adequate error performance for T1 spans is obtained by providing repeaters every 6,000 feet (1,828 meters). This particular distance is important because it was the spacing for the amplifiers for the previously used analog carrier system which T1 replaces. Thus, the same manholes can be used when a system is replaced. The energy in the DS-1 signal is concentrated at 772 kHz, which gives a loss figure of 26.6 dB at 6,000 feet over 22-gauge paired wire cables.

Received Signal

The receiver stays in frame alignment with the transmitted signal by using a clock recovery timing circuit.

It is the job of the receiver at the end of the digital transmission to receive and amplify the transmitted signal and convert it to the original voice signal. In order to accomplish this task, the receiver must stay in frame alignment with the transmitted signal. In order to stay in frame alignment, it must extract its information from the transmitted signal and must operate with the same timing as the transmitter. It does this with a circuit very similar to the clock recovery timing circuit in the repeater.

A rectifier, differentiator, and clipper form trigger pulses that feed a similar tuned circuit to produce a constant frequency timing signal. Square wave clock pulses are produced from the ringing signal with limiters; timing pulses are generated at the zero crossings of the square wave to produce accurate timing at the receiver. With accurate timing, the receiver is able to search and check to make sure it stays in frame alignment.

Frame Alignment

Frame alignment is achieved by a constant vigil for the frame alignment word (FAW), thereby keeping the receiver in synchronization.

Figure 8-14 shows a state diagram that describes how the receiver stays in frame alignment. As shown in *Figure 8-14b*, there is a word in the frame designated as a frame alignment word (FAW). If the receiver is out of frame alignment—state 5 of *Figure 8-14a*—the receiver begins a search for the FAW as shown in *Figure 8-14b*. If the FAW is detected in two successive frames, alignment is assumed and succeeding FAW positions are checked to verify alignment. This is the progression from state 5 to state 6 to state 7 to state 1 of *Figure 8-14a*. From state 1, if there is a slip, the receiver checks for synchronization in state 2, then state 3, then state 4. If a FAW is found in any one of these states, the receiver immediately is back in synchronization. Note that if the system is initially in alignment but slips, the test for a FAW must fail three times before it is declared to be out of alignment, but with only one find of a FAW, the system is declared back in synchronization.

Figure 8-14.
Frame Alignment
(Courtesy of McGraw-Hill Book Co.)

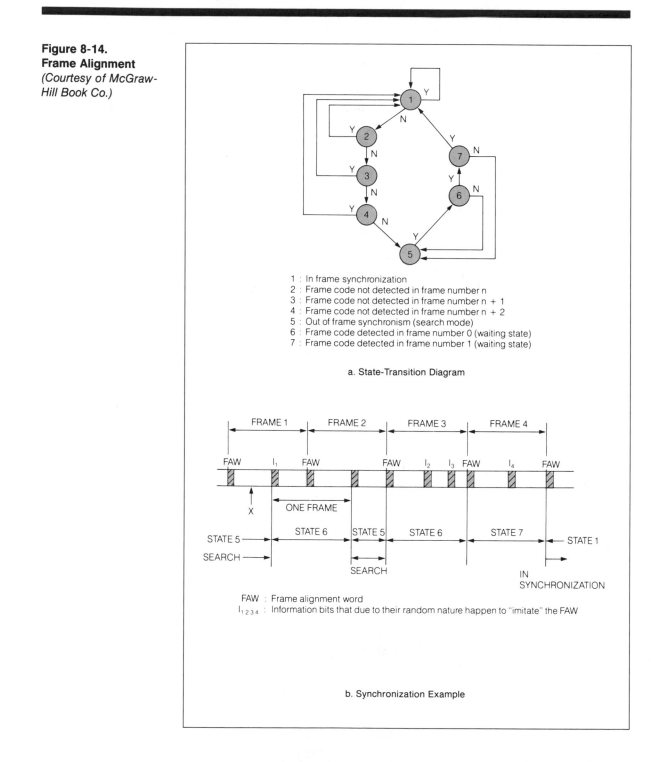

1 : In frame synchronization
2 : Frame code not detected in frame number n
3 : Frame code not detected in frame number n + 1
4 : Frame code not detected in frame number n + 2
5 : Out of frame synchronism (search mode)
6 : Frame code detected in frame number 0 (waiting state)
7 : Frame code detected in frame number 1 (waiting state)

a. State-Transition Diagram

FAW : Frame alignment word
I_{1234} : Information bits that due to their random nature happen to "imitate" the FAW

b. Synchronization Example

MORE LINE CODES

Unfortunately, during transmission of standard AMI signals, the pulse density is not adequate for clock recovery without any slips. Special coding techniques have been used to overcome this problem. For example, bits are added to words that contain all 0s. Or, as is recommended by CCITT for Europe for the first order multiplexer, the even-numbered bits within each 8-bit word are inverted prior to multiplexing. Both methods ensure that enough 1s are in the bit stream for good clock recovery. Let's look at several of these line codes designed specifically to aid clock recovery. *Table 8-2* lists some of the ones used in multiplexing in North America, their bit rates, tolerance, and the type of line code. The ones of most interest are the B6ZS and the B3ZS.

**Table 8-2.
AT&T DS Series Line
Rates, Tolerances and
Line Codes (Format)**

Signal	Repetition Rate (Mbps)	Tolerance (ppm)[a]	Format	Pulse Duty Cycle(%)
DS-0	0.064	[b]	Bipolar	100
DS-1	1.544	±130	Bipolar	50
DS-1C	3.152	± 30	Bipolar	50
DS-2	6.312	± 30	B6ZS	50
DS-3	44.736	± 20	B3ZS	50
DS-4	274.176	± 10	Polar	100

[a]Parts per million
[b]Expressed in terms of slip rate

Binary N-Zero Substitution (BNZS)

To maintain accurate timing, the binary N-Zero substitution (BNZS) assures that there are sufficient 1s in a data stream, by substituting a certain predetermined "mostly 1s pattern" algorithm as replacement for a long string of 0s.

The technique used to solve the pulse density problem is to use special coding to assure that sufficient 1s are always available to provide accurate timing. The special code patterns are inserted into the transmitted bit stream. At the receiver, the presence of the special code is detected and the signal adjusted to the original information.

One way of assuring that there are large numbers of ones in any transmitted bit stream is to detect long strings of zeros and substitute for them code patterns containing mostly ones. At the receiver these code patterns are recognized and the inserted pattern is replaced with 0s. Such algorithms are called Binary N-Zero Substitution schemes or BNZS.

B6ZS and B3ZS Codes

To insert the added 1s for accurate timing, bipolar transmission violations trigger the necessary substitutions.

One special coding used in the Bell System's T2 transmission lines (the DS-2 signal) is called Binary 6-Zero Substitution (B6ZS). In this special coding, the receiver detects where the substitution has occurred by detecting a bipolar violation. Bipolar violations are illustrated in *Figure 8-15*.

**Figure 8-15.
Examples of Bipolar
Violation**

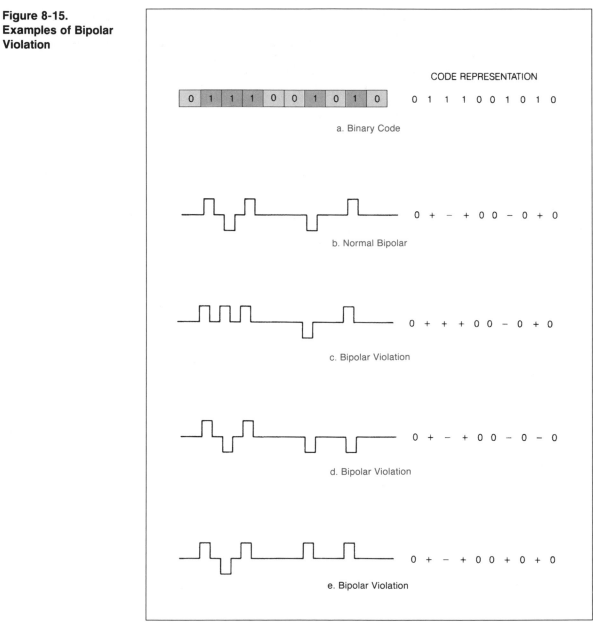

In a normal bipolar code, shown in *Figure 8-15b*, the pulses representing 1s are alternately positive and negative around zero. Zcro is represented by 0 and 1 is represented by a + or − sign which indicates the polarity of the pulse. For the normal bipolar code, the + and − signs alternate in the code with 0s placed as they occur in the input stream.

Bipolar violations are detected when the alternating pattern of + and − signs is interrupted; that is, if two + or two − signs occur in sequence. Three examples of violations are shown in *Figures 8-15c*, *d*, and *e*.

In the B6ZS code, whenever a string of six zeros occurs, a special code shown in *Table 8-3* is substituted. The code pattern substituted depends on the polarity of the 1 pulse immediately preceding the string of six zeros. Note also that in the substituted code, there is a built-in bipolar violation caused by the substitution of the second and fifth pulses. When the second pulse is substituted it causes a +0+ or a −0− violation, and when the fifth pulse is substituted it causes a −0− or a +0+ violation. Here are specific examples:

1. The 1 pulse preceding six zeros is a +
   Binary Code: 0 1 0 1 <u>0 0 0 0 0 0</u> 1 1 <u>0 0 0 0 0 0</u> 1 0 0 1
   Representation 0 − 0 + <u>0 + − 0 − +</u> − + <u>0 + − 0 − +</u> − 0 0 +
   with Substitution:

2. The 1 pulse preceding six zeros is a −
   Binary Code: 0 1 1 1 <u>0 0 0 0 0 0</u> 0 0 <u>0 0 0 0 0 0</u> 1 1 0 1 0 0 1
   Representation 0 − + − <u>0 − + 0 + −</u> 0 − + <u>0 + −</u> + − + + 0 0 −
   with Substitution:

1 Pulse Polarity	Special Code
+	0 + − 0 − +
−	0 − + 0 + −

HDB3

The HDB3 code replaced four consecutive zeros with the codes shown.

A somewhat different special code substitution is recommended by the CCITT in Europe. It is also a BNZS code called High-Density Bipolar Three-Bit substitution (HDB3). This technique replaces strings of four zeros with the substitute code shown in *Table 8-4*.

Table 8-4.
HDB3 Rules

1 Pulse Polarity	*Special Code Odd	Even
+	0 0 0 +	− 0 0 −
−	0 0 0 −	+ 0 0 +

Code Substitution governed by the number of 1s that have occurred since the last substitution.

Because four zeros are replaced (greater than 3), the code is called HDB3. Its substitution rules depend on how many 1s have occurred in the PCM bit stream since the last substitution as well as the polarity of the 1 pulse that is just ahead of the string of zeros. Here are examples:

A. Plus, odd; minus, odd
Binary Code: 0 1 0 1 1 <u>0 0 0 0</u> 1 0 1 0 1 <u>0 0 0 0</u>
Substitution: 0 + 0 − + <u>0 0 0 +</u> − 0 + 0 − <u>0 0 0 −</u>

B. Plus, even; plus, odd
Binary Code: 0 1 0 1 <u>0 0 0 0</u> 0 1 0 1 1 <u>0 0 0 0</u>
Substitution: 0 − 0 + <u>− 0 0 −</u> 0 + 0 − + <u>0 0 0 +</u>

Note that a bipolar violation occurs in the last bit of the substituted code.

B8ZS Code

A newer BNZS code, known as B8ZS has been specified by the CCITT for use on T1 carrier systems. It is similar in principle to the earlier B3ZS and B6ZS codes, but in this case, a series of eight consecutive zeros in the transmitting data stream will be replaced by a special code sequence shown in *Table 8-5*. As the table indicates, the code to be substituted will depend on the polarity of the preceding 1. The new code also introduces a bipolar violation which the receiver will recognize in order to restore the original logic states to the data stream.

Table 8-5. B8ZS Rules

1 Pulse Polarity	Special Code
+	0 0 0 + − 0 − +
−	0 0 0 − + 0 + −

1. The pulse preceding eight zeros is a +
Binary Code   . . . 1 0 <u>0 0 0 0 0 0 0</u> . . .
Substitution   . . . + 0 0 0 + − 0 − + . . .

2. The pulse preceding eight zeros is a −
Binary Code   . . . 1 0 <u>0 0 0 0 0 0 0</u> . . .
Substitution   . . . − 0 0 0 − + 0 + − . . .

Manchester Coding

Manchester coding uses square pulse phase to indicate the two states. This method has excellent timing characteristics, little dc wander, and provides error indications.

The systems previously discussed use repeating code patterns to provide adequate pulse densities for good timing recovery and freedom from dc wander, and to provide error indications. A code that has these benefits, but uses only two levels for binary data is the Digital Biphase, or Diphase, or Manchester code. It uses the phase of a square wave signal to indicate a 1 or a 0 as shown in *Figure 8-16b* and *c*. A 0 has an opposite phase waveform from a 1. Every signaling interval contains a zero crossing to provide a good reference for timing recovery and every interval contains an equal amount of positive and negative level for no dc wander.

Figure 8-16.
Manchester Code

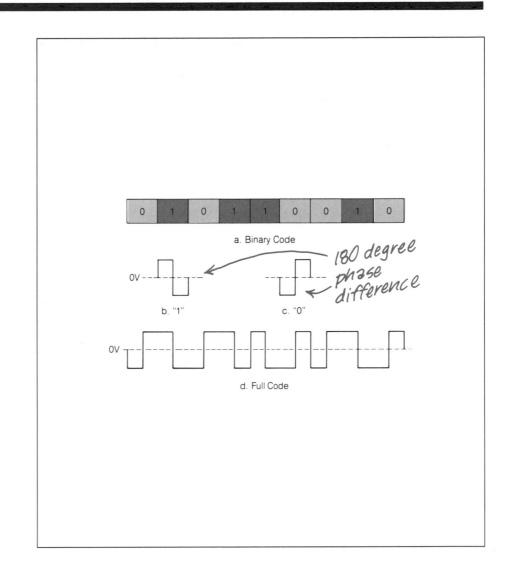

a. Binary Code

b. "1"   c. "0"

180 degree phase difference

d. Full Code

WHAT HAVE WE LEARNED?

1. Digital channel banks produce the PCM signal and multiplex it with many other channels onto a common transmission link.
2. Multiplexers serially arrange PCM signals in time slots with a set number of channels per frame.
3. Multiplexers can have from 24 channels per frame to 4,032 channels per frame and have bit rates from 64 kbps to 274.176 Mbps.
4. Regenerative repeaters retrieve the PCM signal and reconstruct the pulses.
5. Long strings of binary 1s or 0s can cause dc wander and signal transmission errors.

6. Receivers must provide clock recovery for data retrieval based upon the received signal.

7. Long strings of 0s in the PCM bit stream can cause a loss of clock and a slip or loss of synchronization.

8. Special line codes are substituted in the transmitted signal to make sure that a proper number of 1s appear in the bit stream to maintain proper timing.

Quiz for Chapter 8

1. TDM carrier systems are growing in number because:
   a. they are easy to install and maintain.
   b. their error rates can be made very low.
   c. they can use low cost large-scale integrated circuits.
   d. they can carry signals other than voice.
   e. all of the above.

2. The noise generated by TDM carrier systems is _____ that generated by analog carrier systems.
   a. greater than
   b. about the same as
   c. a little less than
   d. much less than

3. One section in a D1 channel bank handles _____ voice channels.
   a. 12
   b. 20
   c. 24
   d. 48
   e. 96

4. Functions performed by D-type channel banks are:
   a. analog to digital conversion.
   b. time division multiplexing.
   c. dialed number translation.
   d. encoding and decoding.
   e. all except c. above.

5. Data sent by a D2 channel bank are at _____ million bits per second.
   a. 1.544
   b. 2.048
   c. 44.736
   d. 1,000

6. Multiplexer systems used in the public network are of the _____ type.
   a. synchronous
   b. plesiochronous
   c. asynchronous
   d. isochronous

7. Pulse stuffing is used in TDMs to:
   a. correct pulses with insufficient height.
   b. add signaling pulses to the data.
   c. provide timing information to the receiver.
   d. equalize the transmission rate between input channels.

8. The Frame Alignment format in the DS-1 signal contains:
   a. 1 bit.
   b. 4 bits.
   c. 7 bits.
   d. 8 bits.

9. Synchronization of digital switches in the North American telephone network is done using a (an) _____ technique.
   a. independent
   b. master-slave
   c. mutual conductance
   d. plesiochronous

10. The line coding scheme used
for the DS-1 signal is:
  a. non-return-to-zero.
  b. binary 3-zero substitution.
  c. ternary.
  d. bipolar.

Modems and Fax Machines— Other Telephone Services

ABOUT THIS CHAPTER

Voice signals—in either analog or digital form—are not the only signals carried over telephone lines. Digital equipment such as computers and facsimile machines routinely send and receive information and control signals through the telephone network.

In order to make use of telephone facilities, digital equipment must be able to translate data into a form suitable for transmission over the network. Equipment at the receiving end will then translate signals back into digital form, check for errors, and make use of the information.

The most common type of equipment used to transmit digital pulses generated by digital devices onto the telephone network is called a *modem*. This chapter will examine the characteristics and performance of modems in detail. It will also introduce the principles and operation of facsimile machines.

WHAT IS A MODEM, AND WHY IS IT REQUIRED?

The modem, which is an acronym for MOdulator-DEModulator, is the interfacing device that couples the output of digital systems to telephone lines. The sending modem converts the digital data to an analog format that can be readily handled by voice grade telephone circuits and the receiving modem changes the format back to recover the original signal.

Computers, like people, must communicate with each other. They pass information back and forth as electrical signals called nonvoice or data. The transmission link for the information varies depending on the physical location. If the machines are located in the same room or the same building, they probably are directly connected. For longer distances, they are connected through the telephone network. Increasing amounts of nonvoice information flow over the public network between machines each day. Let's look at this connection more closely.

Figure 9-1 shows two pieces of digital equipment (data terminal equipment) interconnected to communicate digital information between them. The piece of equipment in *Figure 9-1a* is sending the digital data to the equipment in *Figure 9-1b* located a long distance away. Communications can occur in the same fashion in the reverse direction. A telephone line is the transmission link between them.

As shown in *Figure 9-1a*, the digital equipment outputs bits in parallel to represent the data. The parallel output is converted to a serial bit stream by the transmitter portion of a Universal Asynchronous Receiver Transmitter (UART) so that the information can be sent in serial form over a single line. The least significant bit (LSB) is sent first and the most significant bit (MSB) is sent last. The serial bit stream can't be transmitted directly, but must be changed to tones that can be sent over the telephone

**Figure 9-1.
Communicating Digital
Data**

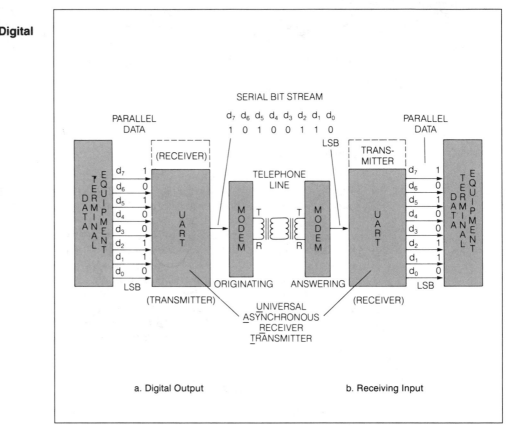

a. Digital Output   b. Receiving Input

line. A modem provides the conversion and coupling to the telephone
line. The word modem is a combination of the words MOdulator and
DEModulator, the two principal functions accomplished by a modem.

At the receiving end of the telephone line shown in *Figure 9-1b*, the
signal is converted back to a serial bit stream by the modem, and the
receiver portion of the UART converts the serial bit stream to parallel data
for input to the digital equipment. Some digital equipment can accept the
serial data directly from the modem; if so, the UART is not needed.

A serial bit stream of 8 bits is shown in *Figure 9-1*. It is input to the
modem at the sending end, called the originating modem, and is output
from the modem at the receiving end, called the answering modem. The
format for each character in the bit stream is shown in *Figure 9-2*.

**Figure 9-2.
Asynchronous (Start-
Stop) Character
Format**

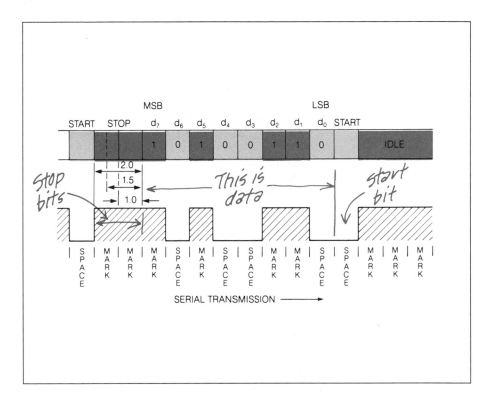

Asynchronous Character Format

The mark and space
intervals are used to
indicate a 1 and 0 re-
spectively. The
mark is also used to in-
dicate "ready for
transmission."

The bit stream has mark and space intervals in the format. A mark is represented by the presence of some predefined voltage or current level, while a space is represented by a different voltage or current level. Since an indication is needed that a transmission link is established and ready even when data are not being transmitted, the idle state is represented by the mark state; therefore, current or tone is present continuously in an idle condition if the circuit is complete. In *Figure 9-2*, a mark represents the binary 1 and a space represents the binary 0 in the digital data as the bit stream is formatted and transmitted.

The asynchronous
format requires there
be a start bit at the
beginning and a stop
bit at the end of the
character code interval
being transmitted.

An asynchronous character format has additional characteristics that provide a timing reference for the receiver. Before the code representing a character of the data (which may be a letter, number, command, or special symbol), there must be a start bit. It is one space interval. After the character code, there must be a stop bit. It varies in length from one interval to two intervals depending on the code and the equipment. The start interval indicates the start of the character code and a stop interval provides the receiver circuits a reference for detection of the beginning of the next character. The format allows the character code itself to be of any practical length, but typically is from 5 to 8 bits long.

In asynchronous operation, the transmitter and receiver operate using independent free-running clocks. Each encoded character is surrounded by start and stop bits. Standards have been established for data rate, bit periods, etc.

To summarize, asynchronous operation requires that the transmitter (sending machine) and the receiver maintain within themselves free-running clocks with the same nominal frequency which is accurate to within certain tolerances (which depend on the maximum data rate to be transmitted). Each encoded symbol, usually called a character or a byte, is preceded by a known state of the input signal called the rest or idle state, followed by a reference or synchronizing signal called a start bit. Besides the data rate, additional parameters upon which the sender and receiver must agree are the length of the signal interval (sometimes called the bit time), the number of signaling intervals per symbol (bits per byte or per character), and the minimum length of the stop bit or idle interval which must precede each new character. Typical stop bit intervals are 1.0, 1.42, 1.5, and 2.0 bit times as previously indicated. A 5-bit code like that shown in *Figure 9-3a* is commonly called a Baudot code, after Emile Baudot who devised the first constant length code for teleprinters in 1874. It used a 1.42 stop interval for a total of 7.42 intervals per character.

5-Bit and 7-Bit Codes

The parity bit is used to check for data errors. Whatever the number of ones making up a word, an extra 1 is added before transmission, where needed, to make the total number of ones an odd or even number. This depends on whether the system is programmed for odd or even parity.

To illustrate what the serial bit stream would look like at the input to the originating modem, the letter S is shown in Figure 9-3 represented in two different codes—the 5-bit CCITT alphabet No. 2 code and the 7-bit American Standard Code for Information Interchange (ASCII). Note that the stop interval is shown as 1.5 for the CCITT code and 2.0 for the ASCII. In many cases, an eighth bit, called the parity bit, is appended to the ASCII to use for detecting errors that may occur in transmission.

Parity may be odd or even in a particular system. If even parity is used, the parity bit is set to a 1 if needed to make the total number of ones in the 8-bit code an even number. Conversely, if odd parity is used, the parity bit is set to a 1 if needed to make the total number of ones in the 8-bit code an odd number.

Asynchronous Operation

In asynchronous operation, the start bit determines synchronization for each character. The receiver's oscillator is synchronized on each character individually, and must not drift beyond the width of a bit.

The communication between the data terminal equipment shown in *Figure 9-1* using the format of *Figure 9-2* is said to be asynchronous. That is, there is no signal within the character format that conveys the timing between transmitter and receiver for each bit. The start bit is used as a synchronizing signal by transmitter and receiver for *each* character, but as the receiver inputs the code, an oscillator in the receiver determines the input timing independently of the data. The timing oscillator frequencies of the transmitter and receiver must stay within certain limits so that the signals at the receiver do not drift beyond the correct sampling points at the receiver.

**Figure 9-3.
5- and 7-Bit
Characters in
Asynchronous Format**

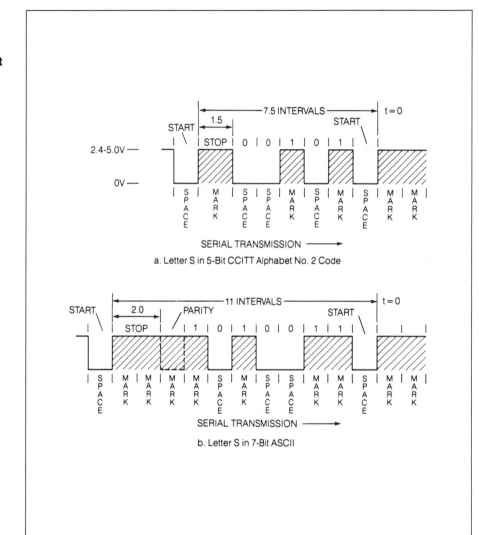

a. Letter S in 5-Bit CCITT Alphabet No. 2 Code

b. Letter S in 7-Bit ASCII

After the first start bit is detected, the sampling rate occurs at the 50% point of each expected bit position. After all 8 bits have been supplied, the detection of the next start bit will adjust the clock, and the process begins again.

Figure 9-4 illustrates the asynchronous operation and shows that even though the transmitter timing clock drifts with respect to the receiver clock, the sampling points of the receiver still can detect the correct bit pattern from the transmitted pulses. The first sample of the data, identified as A in *Figure 9-4*, detects the beginning of the start interval. A timing circuit in the receiver, set for 50% of the average clock time of the transmitter, samples the incoming data again at sampling point B. If sampling point B continues to detect a space level, then the start bit is considered valid and the receiver continues to sample to accept the character bits. Every interval is sampled at the 50% interval point. The timing is determined by the receiver clock. Sampling points C through J

In synchronous operation, a special coded signal is transmitted to keep the receiver oscillator in step with the transmitted data. Data are sent in large blocks between sync information.

recover the 7-bit character code and the parity bit, and samples K and L recover the stop interval. The interval between characters may be of any length. When the next start bit is detected, the timing clocks are reset to synchronize the timing and the sampling process repeats.

Synchronous Operation

Synchronous operation is truly synchronous because a clock signal is transmitted with the data to maintain the transmitter and receiver in continuous synchronization. The format for the digital data is as shown in *Figure 9-5*. Each of the blocks in *Figure 9-5* represents an 8-bit character because a 7-bit ASCII character with parity was selected for the example.

**Figure 9-4.
Asynchronous
Operation of
Transmitter and
Receiver**

Clocks are reset after every word.

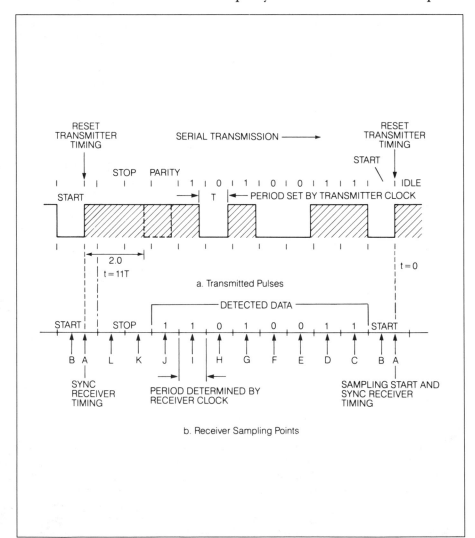

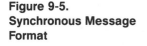

**Figure 9-5.
Synchronous Message
Format**

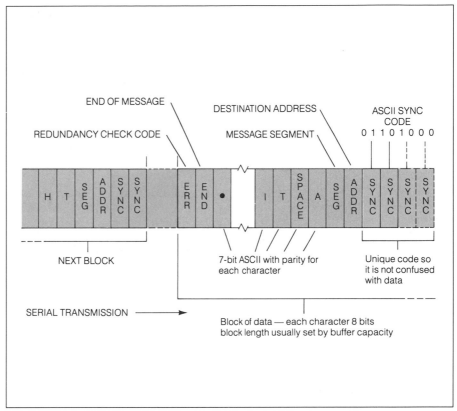

There are no start or stop bits associated with each synchronous character. All the bits for a group of characters are sent one after the other in what are called "blocks of data." As a result, the timing of transmitter and receiver must be more accurately synchronized than for the asynchronous transmission.

After sync codes are detected, data follows until an end of message is detected.

To accomplish this synchronization, special codes are included at the beginning of each block of data. The electronic circuits in the receiver constantly check the incoming data for this pattern. When it is detected, the receiver assumes that the next character is data, and that data will continue until an end of message code is detected. (Because the receiver clocks must be derived from the input data stream, synchronous modems are generally more expensive than asynchronous modems.)

Buffer memories are required both at the transmitter and receiver to store blocks of data before they are transmitted and after they are received.

There are generally multiple synchronization codes at the beginning of each message in case the first code is lost because of transmission problems. The length of the block usually is dependent on a buffer memory that stores the block of data before transmission. This buffer is not necessary in asynchronous transmission because each character is transmitted as soon as it is produced by the digital equipment. But for synchronous transmission, several characters are stored, then all are transmitted at a regular and constant rate of so many bits per second. For the same reason, buffers usually are required at the receiving end.

The synchronous data format shown in *Figure 9-5* is not the only one possible, but it demonstrates the principle. Some synchronization techniques have sync codes inserted at particular time intervals. Some have special framing formats. With the advent of high-speed integrated circuits and very large-scale integration (VLSI), some of the synchronization techniques are changing. More recent circuit designs detect the transitions of the received digital data and continually adjust the receiver clock for even more accurate synchronization.

Isochronous Operation

Isochronous operation is a mix between asynchronous and synchronous operation. *Figure 9-6* shows the character format. Individual characters are framed with a start and stop bit as in asynchronous operation, but the intervals between characters are time controlled. The time interval may be of any length so long as it is restricted to multiples of one character time.

MODULATION AND DEMODULATION

Why is it that the digital signal output from a computer cannot be connected directly to a telephone line? The reason basically has to do with the amount of bandwidth available to carry signals over a single telephone channel, which in turn is related to the cost of the channel. The lowest-cost channels currently available are voiceband channels, which have been designed to carry voice signals produced by telephone sets, and whose bandwidth extends from about 300 Hz to about 3,400 Hz. The means the

In isochronous operation, the characters are each set off by start and stop bits and the intervals between characters are restricted to multiples of one character time.

The voice frequency bandwidth of the telephone channel is inadequate to handle the digital signals outputted by computers.

**Figure 9-6.
Isochronous Character
Format**

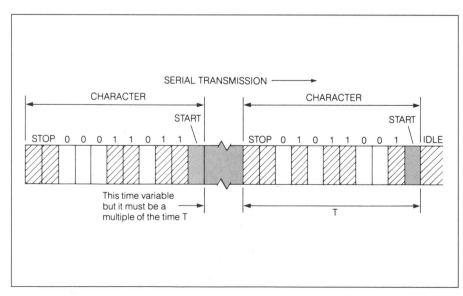

channel cannot pass signals of very low frequency (like direct current), nor signals of very high frequency (above 3,400 Hz). The signals used in computers are usually unipolar signals and the change (transition) from the 0 level to the 1 level is very fast (at a high frequency). As a result, the computer signals contain significant frequencies below 300 Hz (even a dc component) and frequencies well above the 3,400 Hz limit of a VF telephone channel. The fact that the channel will not carry these frequencies leads to the necessity of transforming the digital waveform to a signal that is compatible with the channel and its bandwidth. The means used for the transformation is called modulation.

Modulation

Modulation changes the characteristics of a signal without changing its information. There are three types of modulation—amplitude, frequency, and phase.

Modulation is the process of changing some property of an electrical wave (called the carrier) in response to some property of another signal (called the modulating signal). In the case of transmission of data signals over the telephone channel, modulation involves changing some property of an alternating current wave carrier of between 300 and 3,400 Hz in response to a binary (1 or 0) signal from a computer. The properties of the carrier that are available to be changed are the amplitude, frequency, and phase as illustrated in *Figure 9-7*. Each of these methods is used in modems. Let's look at them in more detail.

Amplitude Modulation

In amplitude modulation, a 1 is represented by a full amplitude carrier level, and a 0 by no amplitude.

Amplitude modulation is the process of changing the amplitude of the carrier in response to the modulating signal. As illustrated in *Figure 9-7b*, when the modulating signal is a binary signal, the amplitude may be varied from 0 for a binary 0 to some maximum value for a binary 1.

Frequency Modulation

In frequency modulation, a 1 is a certain carrier frequency and a 0 is another.

Another property of a carrier that can be varied by a modulating signal is the frequency. *Figure 9-7c* shows how shifting the carrier frequency to a lower value represents a binary zero, and to a higher value represents a binary one. This technique is sometimes called Frequency Shift Keying (FSK), and when the frequencies used are in the voice band, the technique is called Audio Frequency Shift Keying (AFSK). The most widely used modems today carry digital signals in the range of 45 to 1,800 bits per second using the AFSK technique.

Phase Modulation

In phase modulation, the waveforms are shifted in time relative to another wave for a 1, and left unchanged for a 0.

A third property of a carrier that can be varied in response to a modulating signal is the phase. Phase can be visualized as the relative relationship of two waveforms at any given time in their cycle. *Figure 9-7d* illustrates that the phase is shifted for every occurrence of a one bit, but the phase is not shifted for a zero bit. This technique is called Phase Shift Keying (PSK). The phase of the signal sent over the transmission medium usually is not measured absolutely, but rather is measured relative to the phase of the wave during the previous bit interval.

Figure 9-7.
Types of Modulation
*(Source: D. Doll, Data
Communications;
Facilities, Networks,
and System Design,
John Wiley & Sons,
1978, Copyright ©
1978, by John Wiley &
Sons, Inc.)*

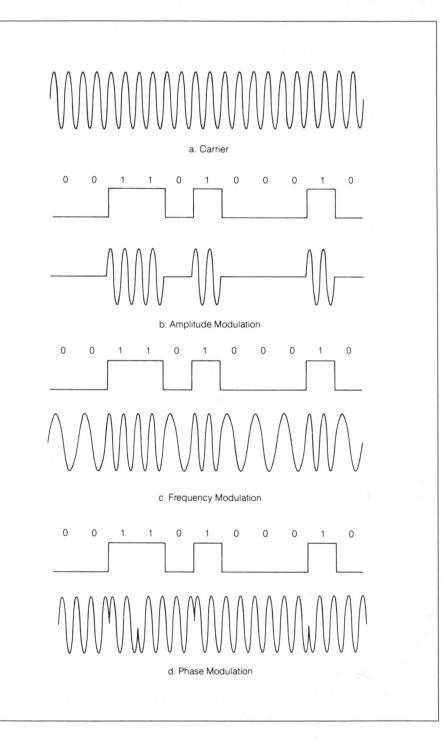

a. Carrier

b. Amplitude Modulation

c. Frequency Modulation

d. Phase Modulation

Transmission of phase information on the telephone network presents some challenges. The human ear is relatively insensitive to the phase of speech or music sounds; therefore, the telephone network was not designed to carefully preserve the phase relationships of signals sent through it. High-speed modems which use phase modulation to carry data usually have circuits in them which compensate for disturbances or nonlinearities in the telephone network to help restore the phase linearity of the received signal.

Demodulation

In demodulation, the telephone signal is processed to recover the original digital stream of 1s and 0s.

As the carrier is modulated to carry the digital signal, so it must be demodulated to recover the digital signal. Detector and filtering circuits sensitive to either amplitude, frequency, or phase recover the 1s and 0s from the modulated carrier signal. Threshold bias and level shifting circuits restore the digital signal to common logic levels or special interface transmission levels specified by the digital signal protocols for interfacing.

ASYNCHRONOUS MODEM OPERATION

A typical low-speed asynchronous modem for half-duplex transmission may use a center frequency (data free) of 1170 Hz, shifting 100 Hz up from center for a logic 1 and 100 Hz down from center for a logic 0. A second center frequency of 2,125 Hz is added for full duplex communications. 2,225 Hz is the 1 frequency and 2,025 is the 0 frequency.

Figure 9-8 shows the frequency modulation schemes used in most low-speed (up to 300 bps) asynchronous modems. In *Figure 9-8a*, a center frequency carrier (1,170 Hz) is frequency shifted to 1,270 Hz for a 1 and to 1,070 Hz for a 0. Every time the serial bit stream for a character is input to the originating modem, the output is a continuous alternating signal of 1,070 Hz or 1,270 Hz depending on whether the input is a 0 or a 1. This scheme is used for simplex transmission (transmission in one direction only) and for half-duplex transmission (transmission in both directions, but only in one at a time).

If full-duplex transmission (transmission in both directions at the same time) is required, then the modulation is accomplished by dividing the available bandwidth into two bands as shown in *Figure 9-8b*. The lower band carries data in one direction and the upper band carries data in the other direction. The lower band center frequency is 1,170 Hz and the frequency is shifted to 1,270 Hz for a 1 and to 1,070 Hz for a 0. The upper band center frequency is 2,125 Hz and the frequency is shifted to 2,225 Hz for a 1 and to 2,025 Hz for a 0. These frequency pairs are the ones used in the Bell System 103 series modem which has become a de facto (by use) U.S. standard.

System Interconnection

When the modem at the transmitter is set in the originate mode and the modem at the receiver is set to the answer mode, the frequencies match up for full duplex operation.

When both modems are operated in the full-duplex mode, *Figure 9-9a* shows the block diagram of the interconnection. The transmitter at the originating modem and the receiver at the answering modem operate at the center frequency of 1,170 Hz, while the transmitter at the answering modem and the receiver at the originating modem operate at the center frequency of 2,125 Hz. *Figure 9-9b* shows the relative signal levels versus frequency for the two signal channels. *Figure 9-9c* shows the frequencies for a similar full-duplex CCITT system used in Europe.

Figure 9-8.
AFSK Modulation

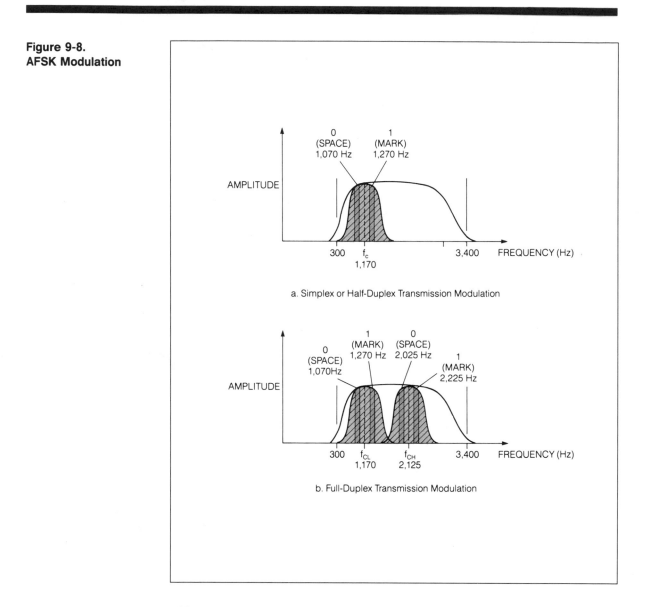

a. Simplex or Half-Duplex Transmission Modulation

b. Full-Duplex Transmission Modulation

1200 bps Asynchroncus Modem

In the Bell 202 modem, the mark and space frequencies are widely spaced because of the increased bandwidth requirements of a higher data speed.

Another method of dividing the available frequencies is illustrated in *Figure 9-10*. In this case, the modem can carry asynchronous data at 1,200 bits per second, but in only one direction at a time (half-duplex operation). The higher data rate requires a wider bandwidth so the two frequencies for 0 and 1 must be spaced further apart; therefore, there is not enough bandwidth in the channel for full-duplex operation. The modems using this scheme are called 202-compatible, since the Bell System 202 modems established this scheme. Enough bandwidth is

Figure 9-9.
Low-Speed
Asynchronous Full-
Duplex Modem
Frequency
Assignments and
Levels *(Courtesy of*
Digital Press, Reprinted
from J.E. McNamara,
Technical Aspects of
Data Communication,
Second Edition,
Copyright © 1982 by
Digital Equipment
Corporation)

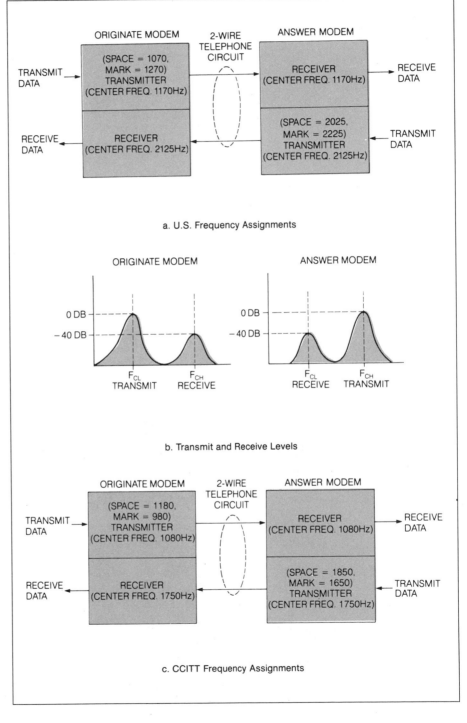

a. U.S. Frequency Assignments

b. Transmit and Receive Levels

c. CCITT Frequency Assignments

**Figure 9-10.
Bell 202 Compatible
Modem**

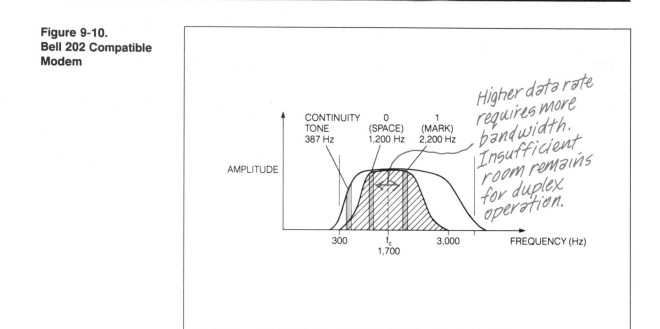

available in the channel with the 202 frequencies to allow a single-frequency signal at 387 Hz to be transmitted in the reverse direction while the 1,200 bps data are being sent forward. This reverse channel is used mainly to transmit a continuity tone for the receiver to tell the transmitter that the circuit is established, but some data may be sent using on-off keying of the tone. Modems of modern design using low-cost electronics to perform more advanced modulation techniques now carry 1,200 bps data in full duplex mode, and some offer compatibility with either 300 bps 103-type signals or 1,200 bps data.

Bell 103-Type Modem

The Bell 103 type modem uses a matching line transformer for coupling.

Figure 9-11 is a block diagram of a 103-type modem. The two-wire telephone line is terminated in a line matching transformer. The transformer secondary is connected to both the receive section input and the transmit section output, but the received signal has no effect on the transmit section. The transmit output signal could affect the receive section, but the receive bandpass filter prevents the transmit signal from entering the receive section because the transmit and receive frequencies are in different bands. The receive bandpass filter also rejects noise and spurious frequencies riding on the receive signal from the telephone line. The limiter eliminates amplitude variations.

Figure 9-11.
Low-Speed
Asynchronous Modem
Block Diagram

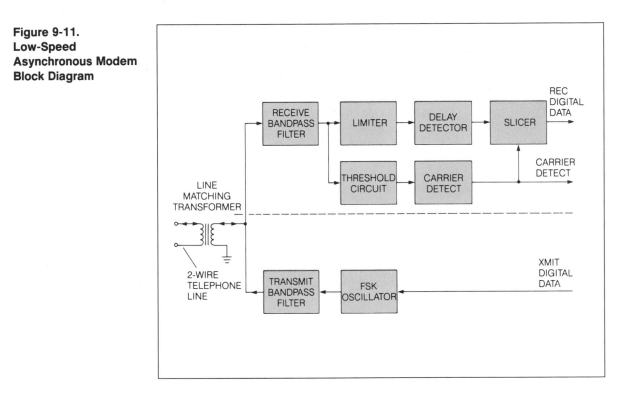

A delay detector provides a delayed sample of the signal, compares it to the receive signal, and gives an output which is proportional to the difference in frequency. The slicer circuit clips the top and bottom of the detected signal and produces a digital signal at the output with the proper digital voltage levels for 1 and 0.

When data are to be transmitted, the digital serial bit stream is applied to a frequency shift oscillator which produces the audio frequency shifted tones representing the ones and zeros. This signal is filtered by the bandpass filter to remove spurious harmonics (particularly those within the receive passband) and is passed through the line matching transformer to the telephone line.

Acoustical Coupler

Acoustic couplers utilize a speaker and microphone set to couple to the telephone handset.

In *Figure 9-11*, the modem is shown to be connected directly by wires to the telephone line. This is the preferred method; however, some modems are acoustically coupled to the telephone handset. The coupler has a microphone and speaker which interfaces with the handset receiver and transmitter, respectively, to couple the transmitted and received signals by sound waves. Thus, there are no wire connections between the modem and telephone line. A telephone call is made to the destination and a connection established the same way as for a conversation before the handset is placed in the coupler.

Because of the sophistication of microelectronics technology, modems have been greatly reduced in size and many new features have been added.

Advances in Modern Modems

Original modems using circuitry like *Figure 9-11* occupied a volume of 600 cubic inches (9,832 cubic centimeters) in a package $10 \times 10 \times 6$ inches ($25.4 \times 25.4 \times 15.2$ centimeters). Using the latest integrated circuits, modems are now manufactured occupying less than one-tenth of that volume, 65.3 cubic inches (1,070 cubic centimeters), in a package $5.5 \times 9.5 \times 1.25$ inches ($14 \times 24.1 \times 3.2$ centimeters). Some modems are built as accessory boards that fit directly into host computers. Besides the smaller physical size, these modern modems provide added performance features.

Here are some examples:

1. Provide data rates up to 9600 bps.

2. Dial telephone numbers automatically for origination and automatically answer incoming calls.

3. Return alphabetic characters to the data terminal equipment to report on the condition of the telephone line and the call in progress.

4. Detect the answer tone from a distant modem automatically and adjust the data rate to match the distant modem's data rate (known as autobaud).

The 10-fold reduction in size and the significant increase in performance and "intelligence" is made possible by using microprocessor techniques and by reducing the electronics of the modem to a single integrated circuit chip.

One Chip IC Modem

The TCM3105E manufactured by Texas Instruments represents a typical one-chip modem which meets Bell 202 and CCITT V.23 requirements. A Frequency Shift Keying approach allows asynchronous operation within the standard telephone voiceband. Data can be modulated and transmitted at 75, 150, 600, and 1200 baud, while signals may be received at 5, 75, 150, 600, and 1200 baud. The TCM3105 can be operated in half-duplex mode up to 1200 baud (transmit or receive only). In full-duplex mode, however, receive data rates are limited to only 150 baud. *Figure 9-12* represents a block diagram for the TCM3105.

Operation

The TCM3105 is made up of four major sections: transmitter, receiver, carrier detector, and control circuit.

A transmitter circuit accepts digital signals from sending digital equipment and converts data into corresponding analog signals which are coupled to a hybrid network as shown in *Figure 9-13*. Transmit baud rate is set by control inputs TXR1 and TXR2. Internal filtering prevents noise and harmonics from being transmitted.

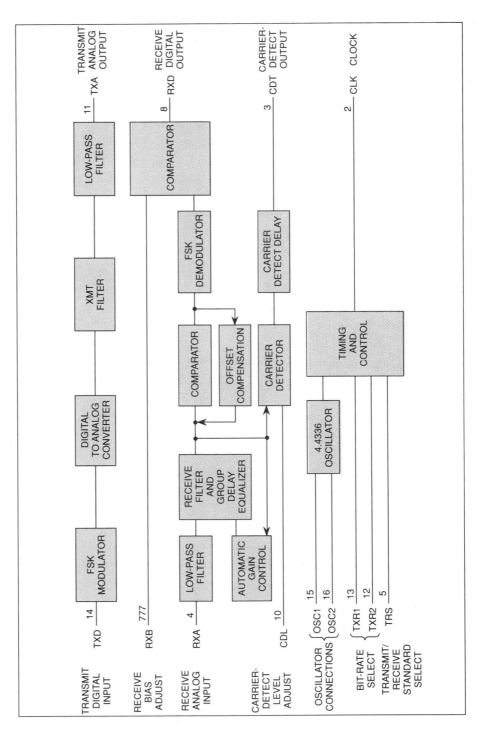

Figure 9-12. TCM3105 Block Diagram. *(Courtesy Texas Instruments, Inc.)*

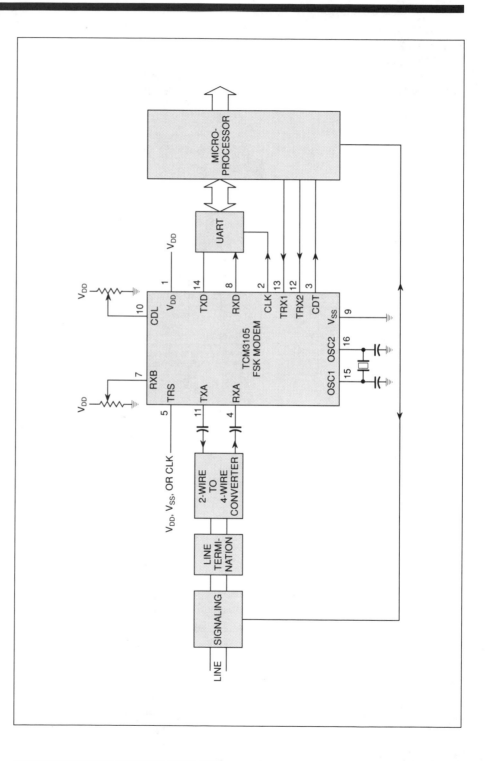

**Figure 9-13.
TCM3105 Typical
Configuration.**
*(Courtesy Texas
Instruments, Inc.)*

The receiver circuitry automatically filters any incoming analog signals and adjusts gain to a proper level. The frequency of the analog signal is converted to a proportional voltage. Since only two frequencies are used in modulation—one for a logic 1 (or "mark") and another for a logic 0 (or "space")—it is a simple matter to convert each frequency into its appropriate logic level. Digital information is then made available to the receiving equipment. Transmit and receive signals are coupled to the telephone line through an interface circuit similar to the network shown in *Figure 9-14*.

A carrier detection circuit compares a reference voltage on the CDL pin with the receive signal output. The CDT pin shows the results of the comparison. A logic 1 at CDT indicates that carrier is present. A logic 0 indicates that carrier is absent. CDT is a very important signal in transmit and receive synchronization. *Figure 9-13* shows this line connected to a microprocessor.

Timing and control circuitry directs the flow of data into and out of the modem by effecting the TCM3105 baud rate. Timing is synchronized by a 4.4336 MHz crystal oscillator connected at the oscillator inputs OSC1 and OSC2.

STANDARDS FOR DIGITAL EQUIPMENT INTERFACE

Several standards have been established for interfacing between digital terminal equipment and modems. The most common U.S. standard is called "EIA RS-232C," and the European standard is called "CCITT Recommendation V.24."

The modem must exchange control signals with whatever terminal or business machine is connected to it. The frightening possibility of large numbers of mismatched control signals and connectors between modems and data terminals has resulted in the establishment of several standards to define the physical and electrical parameters between modems, called data communications equipment (DCE), and terminals, called data terminal equipment (DTE). The most common U.S. standard, administered by the Electronic Industries Association, is called EIA RS-232C. The European standard, administered by the CCITT, is known as CCITT Recommendation V.24.

The basic signals required for sending and receiving data, controlling the modem from a terminal, and passing status information back to the terminal from the modem, along with their standard designations and pin assignments for the standard connector are shown in *Table 9-1*.

Synchronous modems require more interface leads than asynchronous modems because the clocking signal for both the send and receive data must be accommodated, and the standards specify interface leads for selecting alternative data rates (in case of excessive error rates) and a signal from the modem indicating a high probability of an error in the received data.

**Figure 9-14.
TCM3105 Line
Interface Circuit.**
*(Courtesy Texas
Instruments, Inc.)*

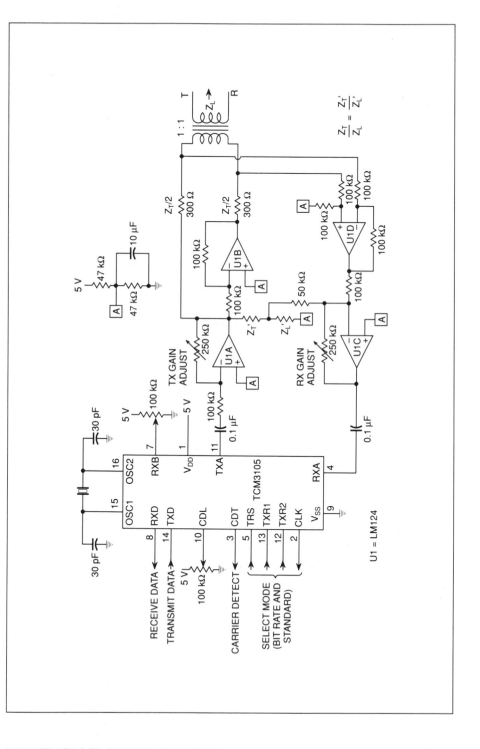

**Table 9-1.
Internationally
Accepted
Assignments for
DTE/DCE Interface
Using a 25-Pin
Connector***

Pin	CCITT	EIA	Circuit Name
1	101	AA	Protective Ground (not always used)
2	103	BA	Transmitted Data
3	104	BB	Received Data
4	105	CA	Request to Send
5	106	CB	Clear to Send
6	107	CC	Data Set Ready
7	102	AB	Signal Ground
8	109	CF	Received Line Signal (Carrier Detector)
9	Often used for modem power test point-do not connect		
10	Often used for modem power test point-do not connect		
11			
12			
13			
14			
15	114	DB	Transmit Signal Element Timing—DCE Source (Synchronous Modems Only)
16			
17	115	DD	Receive Signal Element Timing—DCE Source (Synchronous Modems Only)
18			
19			
20	108	CD	Data Terminal Ready
21			
22	125	CE	Ring/Calling Indicator
23			
24			
25			

(Courtesy of Digital Press, Reprinted from J.E. McNamara, Technical Aspects of Data Communication, Second Edition, Copyright © 1982 by Digital Equipment Corporation)

The interface standards specify pin functions, voltage levels for data and control signals, hardware safety requirements, load parameters, frequency response, and crosstalk.

In addition to specifying the standard pin designations and connection, the EIA RS-232C/CCITT V.24 standards specify signal conditions and levels that must occur on the interchange. For example, the voltages specified by the EIA standard are +3 V to +25 V for a space (zero) and −3 V to −25 V for a mark (one). The circuit implementation must meet certain electrical specifications other than voltage, such as withstanding dead shorts between conductors without damage, and cannot exceed maximum values for load resistance and effective capacitance. To aid in meeting the line interface requirements, special line drivers and line receivers have been designed in integrated circuit form. If a line is to receive data that are being sent under RS-232 standards, a SN75189 line receiver can be placed in the line. If a line is to transmit signals at RS-232 standards, then a SN75188 line driver can be used to drive the line. The interconnection is shown in *Figure 9-15a*.

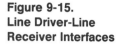

**Figure 9-15.
Line Driver-Line
Receiver Interfaces**

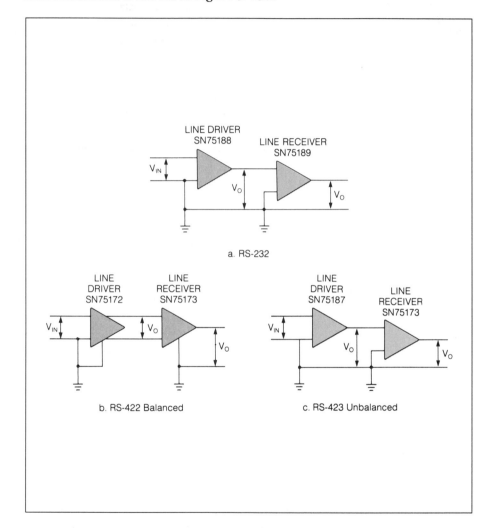

The EIA RS-232/CCITT V.24 standards specify an unbalanced or single-ended interface for each control circuit (*Figure 9-15a*). Such circuits have problems with noise immunity and ability to drive long lengths of cable. Other standards, designated EIA RS-422 and EIA RS-423 specify both a balanced and unbalanced interface as shown in *Figure 9-15b* and *c*. In addition to different termination configurations, these standards specify limits for the rise time of the digital signals to lower the probability of crosstalk on adjacent circuits and improve the maximum signaling distance from 15 meters to over 1,000 meters for data rates under 100,000 bits per second. New EIA standards RS-485 that are very similar to RS-422 cover the case where up to 32 drivers and 32 receivers may be on the same line.

ERROR DETECTION AND CORRECTION

Error control is built into a system by using microprocessors to manage the redundant transmission of data. By comparison of the two sets of data, the data is either accepted as perfect or if the error is found, a request is sent to retransmit.

Although the subject of error detection and correction has nothing to do directly with the subject of modems, whenever digital data are transmitted over telephone channels using modems, errors are sure to occur. The computer systems and terminals that produce and consume the data must take measures to detect and, if possible, correct the data errors.

Recent modem developments combined with the use of microprocessors have allowed the data to be buffered, error control applied, and checked by the receiver. If the receiver detects an error, it automatically requests retransmission of the erroneous data.

Error control is applied by adding redundancy; that is, information in addition to the minimum required to send the original data. The redundant information is related to the original input in some systematic way so that it can be regenerated when it is received. Upon receipt, if the regenerated error control matches that which was sent along with the data, it is assumed the transmission is error free.

Errors may be detected by using the cycling redundancy check (CRC) code. One method involves dividing the data by a mathematical expression at the transmitter site and transmitting the remainder portion of the answer. At the receiving end, the same mathematical manipulation of the data is duplicated and the two remainders compared.

The required redundancy is provided in different ways. The redundancy to detect errors in long blocks of data is provided by a class of codes called cyclic redundancy check (CRC) codes. The process of generating a CRC for a message involves dividing the message by a polynomial, producing a quotient and a remainder. The remainder which usually is two characters (16 bits) in length (see ERR of *Figure 9-5*) is added to the message and transmitted. The added information is sometimes referred to as a block check character (BCC). The receiver performs the same operation on the received message and compares its calculated remainder with the received remainder. If they are equal, the probability is quite high that the message was received correctly.

ADVANCED STANDARDS

Digital communication is constantly pushing the limits of technology by trying to carry ever-increasing amounts of information over the conventional telephone network. As long as the transmission medium offers enough bandwidth to support the desired data rate, data transfer can be achieved using conventional modem techniques. V.22 is a common standard used to define the operation of 1200 bps modems. Data rates as high as 2400 bps can be implemented with the V.22bis standard. Under V.22bis (the suffix "bis" indicates that the standard is a secondary or offshoot of an already existing standard), data is organized into 4-bit words, then sent at 600 baud. This supports full-duplex communication between two modems using regular asynchronous modem techniques covered earlier in this chapter.

At speeds above 2400 bps, the regular telephone network simply does not offer enough bandwidth to carry the additional information.

Other methods had to be developed. Until about 1984, however, any modem transmissions above 2400 bps required four-wire leased telephone lines for more bandwidth and greater noise immunity than conventional two-wire telephone lines. Although standards like Bell 208 for 4800 bps, V.29 for 9600 bps, and V.33 for 14.4 kbps were available at that time, the modem equipment was complex, and leased lines were (and are) expensive. Only large corporations needing to transfer huge amounts of data could justify the expense.

High-speed Modems

9600 bps modems are available that operate over the standard telephone network.

One of the greatest breakthroughs in modem technology came in 1984 with the acceptance of the V.32 standard supporting full-duplex 9600 bps over the general switched telephone network. As V.32 modems became available in 1986, everyday computer users now had access to high-speed communication.

The technique of 9600 bps data communication is more involved than that for slower modems. Since a single 3 kHz telephone voice channel can carry one 2400-baud data signal, a V.32 modem places data into 4-bit words and transmits at 2400 baud. Both modems then must transmit simultaneously and sort out their own transmitted signal from any received signal. Echo cancelers are used in each V.32 modem to provide this isolation.

Echo cancellation removes reflected transmissions from any received signal to allow two high-speed modems to operate simultaneously in full-duplex mode.

Echo cancelers take advantage of the fact that telephones are never ideally matched to the transmission line—that is, the impedance of the telephone circuit is never exactly equal to the impedance of the telephone (or modem) hybrid. As a result, a small amount of the transmitted signal is reflected from the destination back to the transmitting end of the communication link. Some transmitted signals are also reflected from the local hybrid and never leave the sending modem. Both of these reflected signals must be subtracted from the total signal. What is left represents the received signal.

Trellis encoding is a coding technique used in V.32 modems to reduce data errors by recognizing known patterns in received signals.

V.32 modems also use an advanced coding technique called trellis encoding which allows consecutive received signals to be examined for known patterns. This technique can substantially reduce errors encountered in data transmission, and has made V.32 a reliable, high-speed technique.

The CCITT is now evaluating a V.32 standard for 14.4 kbps (dubbed V.32bis) which requires extremely good echo cancelers and improved receiver circuits. Preliminary tests have proved very promising and V.32bis is expected to be formally adopted by the CCITT sometime in 1991.

Error Detection and Correction

As modems achieve higher speeds, the probability of data errors increases. This is due to the lower signal levels used in high-speed modems, as well as the effects of noise at the edges of bandwidth. V.42 was approved in 1988 as a standard error correction technique using advanced cyclical redundancy checks and the principle of Automatic Repeat Request (ARQ).

The V.42 standard
incorporates an Auto-
matic Repeat Request
signal (ARQ) into
its data correction
technique.

To use ARQ, transmitted data is grouped into blocks and cyclical
redundancy calculations add error checking words to the transmitted data
stream. Since the modem itself supplies the error checking information,
data transfer is much faster. The receiving modem calculates new error
check information for the data block and compares it to the received error
check information. If the two codes match, received data is valid and
another transfer takes place. If the codes do not match, an error has
occurred and the receiving modem requests a repeat of the last data block.
This repeat cycle will continue until valid data is received.

V.22 modems began to appear with V.42 error correction in 1989.
V.32 modems now commonly incorporate V.42 in their designs.

Data Compression

The V.42bis data
compression standard
is based on the Lempel-
Ziv mathematical algo-
rithm and is used by
the modem itself. No
pre-compression is
required by the host
computer. Compres-
sion can boost effective
data rates as high as
34.8 kbps in V.32
modems.

Even higher data rates can be achieved by using data compression
techniques. V.42bis was approved in 1989 as the first official modem data
compression standard. V.42bis modems look for repeated or common
patterns of transmitted data and insert shorter pieces of data to represent
it. A receiving modem will reconstruct the original data from the shorter
code segment. Real-time compression and decompression is provided by
the modem itself using a variation of the Lempel-Ziv algorithm. This
means that effective data rates as high as 34.8 kbps can be supported with
V.32 technology.

The effectiveness of data compression depends on the data to be
encoded. Highly repetitious data will be compressed well, while
nonrepetitious or pre-encrypted data may be compressed little, if at all.

V.42bis data compression is now appearing in V.22bis and V.32
modems. As V.32bis modems reach the market, they will probably be
equipped with V.42bis as well.

Other Data Compression Techniques

Data compression is widespread in the digital network because it
offers economy. Compression reduces the number of characters in a
transmission. Not only does this reduce the probability of errors, but the
time required to send the data can be significantly reduced. Lower
"connect time" will lower the cost of sending data. There are over 100
algorithms in use that can compress data for transmission. Algorithms can
generally be divided into character-encoding and statistical-encoding
techniques.

Character-encoding techniques examine a data stream one character
at a time, then replace common or repeated character strings with shorter
ones (Lempel-Ziv is just one example of this approach). The receiver
recognizes compressed strings and regenerates the original string. *Run
length compression* is a popular format that is often used to compress
computer graphic images for storage on mass media devices. A string of
four or more repeating characters are counted and replaced with a three
character code.

Statistical encoding calculates the probabilities of occurrence for
both individual characters and character patterns. Common patterns can

be replaced with short codes while less-common patterns will require longer code sequences. The *Huffman* encoding scheme operates just this way. Huffman coding produces very short words that are easy to decipher without being mistaken for a prefix of another word.

The number of bits needed to represent a character using Huffman compression is often expressed as:

$$B = \text{Integer}(-\log_2 P)$$

where

    B   is the actual number of bits,
    P   represents the probability of occurrence.

The term "Integer" in the equation indicates that the number must be rounded to its nearest whole number. For example, if the letter "E" has a probability of 0.13 (13%) in normal text, B would then be [Integer(2.94)] 3, so 3 bits would be needed to represent E instead of 7 or 8 ASCII bits. A variation of Huffman encoding is used to compress run length data developed for transmission on facsimile machines.

PROTOCOLS

Protocols are sets of rules for communicating between digital equipment that are implemented by software techniques rather than hardware.

There are also rules for the interaction of communications equipment that usually are implemented through the programming of the data terminal equipment involved rather than being built into the hardware. These are called *protocols*. Protocols set the rules for grouping bits and characters (framing), error detection and correction (error control), the numbering of messages (sequencing), separating control and data characters (transparency), sorting out receiving and sending equipment (line control), the actions required on start-up (start-up control), and the actions required on shut-down (time-out control). Protocols may be character oriented like IBM's Binary Synchronous Data Transmission (BiSync); they may be byte oriented like Digital Equipment Corporation's Digital Data Communication Message Protocol (DDCMP), or they may be bit oriented like IBM's Synchronous Data Link Control (SDLC).

An example of each of the three protocols is shown in *Figure 9-16*. The character oriented BiSync protocol uses different characters in the character set to indicate the beginning of the heading, the beginning of the message text, and the end of the text. The character indicating the beginning of the message header, called the SOH, causes the accumulation of the redundancy check character (BCC) to be reset. The byte oriented DDCMP protocol begins with a header that gives the type of message, the number of text characters in the message, a message number, destination address, and a Cyclic Redundancy Check (CRC). The message text follows and the frame is ended by another CRC. The frame of the bit oriented SDLC protocol begins and ends with the unique bit sequence 01111110. The bits shown in the Frame Check Sequence (FCS) are the CRC of the frame. The most important property is that the message is always transparent; that is, it may contain any sequence of bits, without being mistaken for a control character.

Figure 9-16.
Protocol Examples

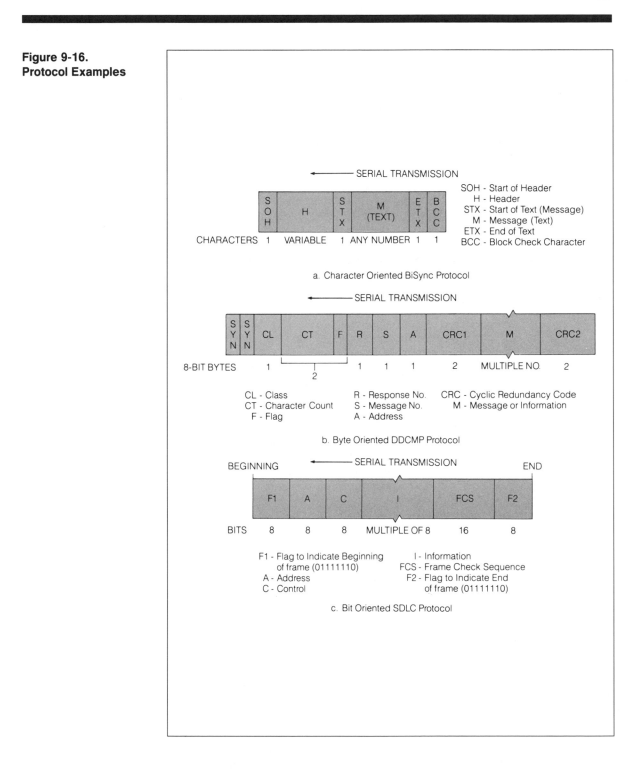

SERIAL TRANSMISSION ←

S O H	H	S T X	M (TEXT)	E T X	B C C

SOH - Start of Header
H - Header
STX - Start of Text (Message)
M - Message (Text)
ETX - End of Text
BCC - Block Check Character

CHARACTERS   1   VARIABLE   1   ANY NUMBER   1   1

a. Character Oriented BiSync Protocol

SERIAL TRANSMISSION ←

S Y N	S Y N	CL	CT	F	R	S	A	CRC1	M	CRC2

8-BIT BYTES   1   CL(CT)2   1   1   1   2   MULTIPLE NO.   2

CL - Class   R - Response No.   CRC - Cyclic Redundancy Code
CT - Character Count S - Message No.   M - Message or Information
F - Flag   A - Address

b. Byte Oriented DDCMP Protocol

BEGINNING   SERIAL TRANSMISSION ←   END

F1	A	C	I	FCS	F2

BITS   8   8   8   MULTIPLE OF 8   16   8

F1 - Flag to Indicate Beginning   I - Information
     of frame (01111110)   FCS - Frame Check Sequence
A - Address   F2 - Flag to Indicate End
C - Control   of frame (01111110)

c. Bit Oriented SDLC Protocol

FACSIMILE

Perhaps the most popular example of a digital communication system is the facsimile (or fax) machine. Few electronic instruments—except maybe the telephone itself—have enjoyed such tremendous acceptance and growth. Simply stated, fax machines digitize printed images and transmit corresponding data over the general telephone network. They also receive digital image information placed on the network and reconstruct the received image on paper. Although modems make up only one part of a fax machine, it still serves as a critical interface between digital equipment and the telephone network. This section of the book will examine the facsimile process.

Figure 9-17 shows a simplified block diagram of a typical, full-function fax machine. In spite of their small size and sleek appearance, fax machines incorporate a surprising amount of electronic and mechanical components into their operation.

Central Microprocessor

A central microprocessor is at the heart of all fax machines. It is responsible for managing all fax operations and coordinating the flow of data into and out of the system. Like any microprocessor, a certain amount memory is needed. Permanent memory (ROM—Read Only Memory) stores a program used to run the fax machine. ROM can also store tables of reference data, characters used in the display, or other such information. Temporary memory (RAM—Random Access Memory) holds results of calculations, variables, system status flags, or any information that will change regularly during normal operation. *Figure 9-18* is a block diagram of a conventional microprocessor/memory configuration. The exact amount of memory will depend on the particular fax machine.

A central microprocessor is used to manage the overall operation of the fax machine.

The Fax Modem

Transmitted or received image data is stored in a special memory buffer to be processed by the modem or printing mechanism.

A microprocessor interface circuit accepts commands from the microprocessor and directs the modem circuit. When transmitting, the modem will accept scanned image data from an image memory buffer, translate data into analog signals, and deliver the information to the telephone network. When receiving, analog tones on the telephone line will be converted to digital information by the modem IC and stored in the image memory buffer for further processing and printing. *Figure 9-19* is a block diagram of a typical fax modem circuit. The modem can also detect incoming calls, as well as provide dial signals.

**Figure 9-17.
Basic Facsimile Block
Diagram**

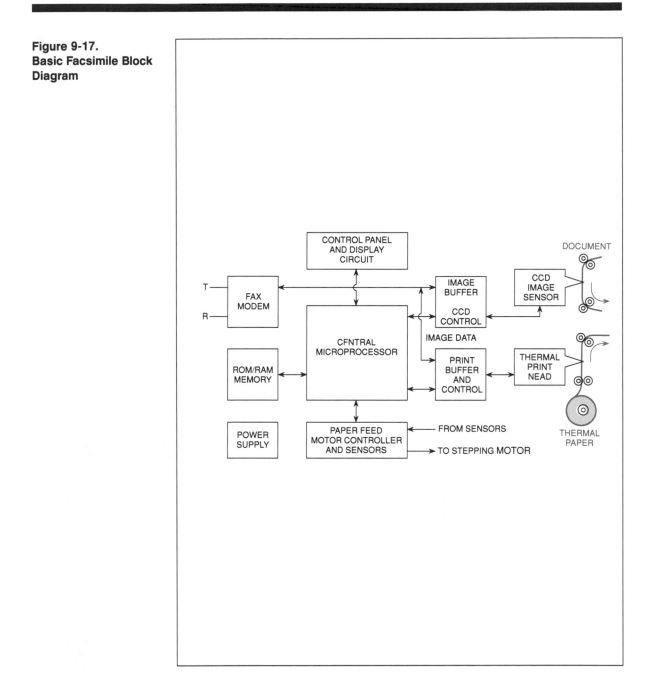

**Figure 9-18.
Microprocessor Block
Diagram**

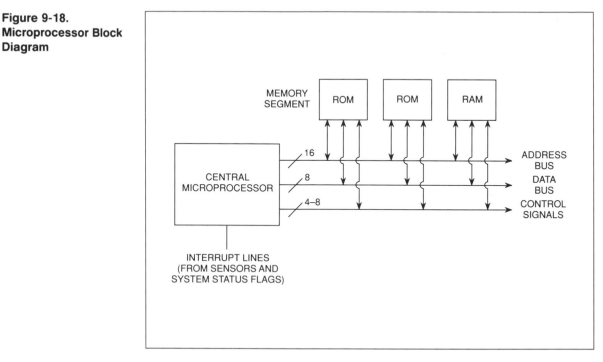

Control Panel

Control panel circuitry serves two primary functions in the fax machine. First, it allows the user to input operating parameters such as date, time, print resolution, baud rate, and desired destination telephone number from the front panel. Second, system status and operating parameters are often displayed on the front panel in the form of a liquid crystal display. An autonomous microprocessor is often used to provide exclusive supervision of the control panel. This remote microprocessor will transfer commands and data to the central microprocessor, as well as receive system status and display information. A simplified control panel is shown in the diagram of *Figure 9-20*.

Receiving

Data enters the fax machine through the modem where it is translated into digital information and interpreted by the central microprocessor. Image data is stored in the image buffer memory. The printing circuit is activated by the central microprocessor. Image data is loaded into the print buffer and processed into signals for the fax machine print mechanism.

**Figure 9-19.
Modem Block Diagram**

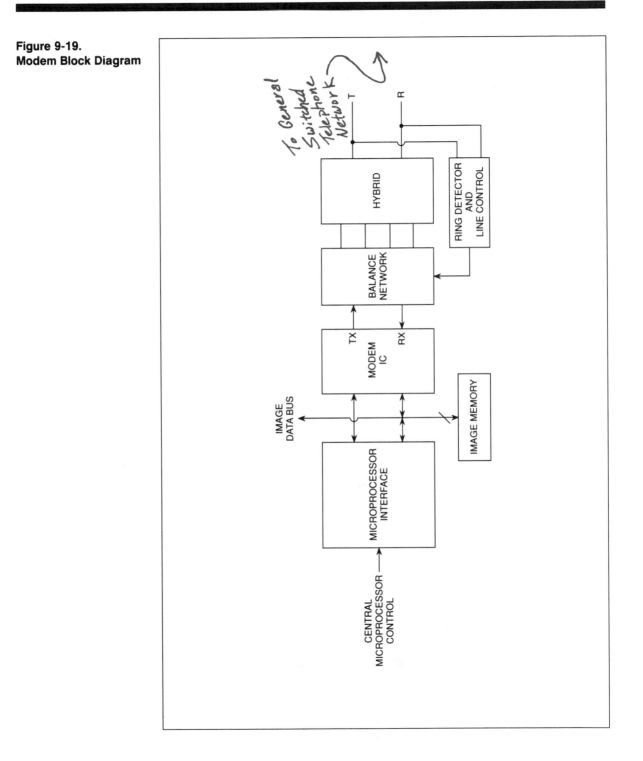

**Figure 9-20.
Control Panel Block
Diagram**

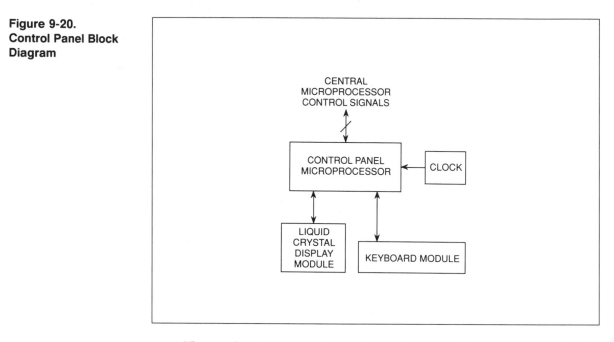

Thermal printing is being replaced by electrostatic printing like the kind used in laser printers. This will allow "plain-paper fax machines."

The predominate printing technique is thermal line printing. That is, the image is printed one line at a time onto thermal paper as indicated by the diagram of *Figure 9-21*. Thermal printing has enjoyed tremendous popularity due to its characteristics of low-noise, low-power consumption, and high-reliability (few moving parts). Thermal paper, however, is flimsy, expensive, and fades with time. Eventually, thermal printing will be replaced with electrostatic printing—the same technique used in laser printers. The electrostatic process allows standard, single-sheet paper to be used. Such devices are called "plain-paper fax machines."

Transmitting

During transmission, the central microprocessor activates the document feeder and line scanning circuits. The document is fed through the fax and scanned one line at a time using CCD (charge coupled device) contact image sensors as shown in *Figure 9-22*. The line sensor delivers a line of pixel data to the scan controller where it is processed and stored as image data in the image buffer memory. The central microprocessor interprets the image data and delivers it to the modem. The modem, in turn, places the data on the telephone network. Both document feed and paper feed are accomplished using stepping motors under the direction of discrete control and driving circuitry.

COMPUTER FAX DEVICES

The ever-increasing use of personal computers has created a need for facsimile functions in the computer itself. A new generation of fax devices has been developed to transmit and receive text and graphics files

**Figure 9-21.
Printing System Block
Diagram**

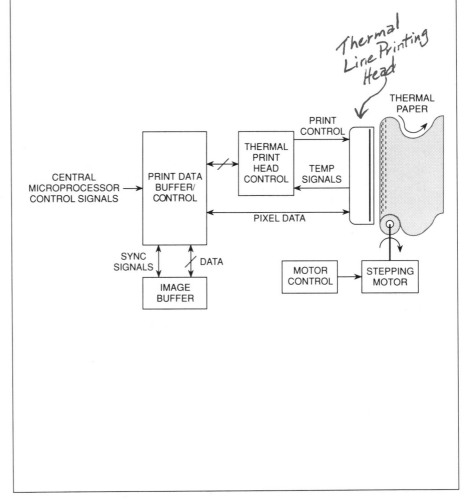

between computers and fax machines (or other computers). This "computer fax" eliminates the need for line scanning and thermal printing, so mechanical components are essentially eliminated. What is required, however, is a computer program that will translate text and graphics into data which is acceptable for fax transmissions, and convert any received data into text/graphics that can be viewed on a computer monitor. Software and single-board modems and fax machines have been developed that can be plugged into expansion slots of computers.

Figure 9-22.
Image Acquisition
Block Diagram

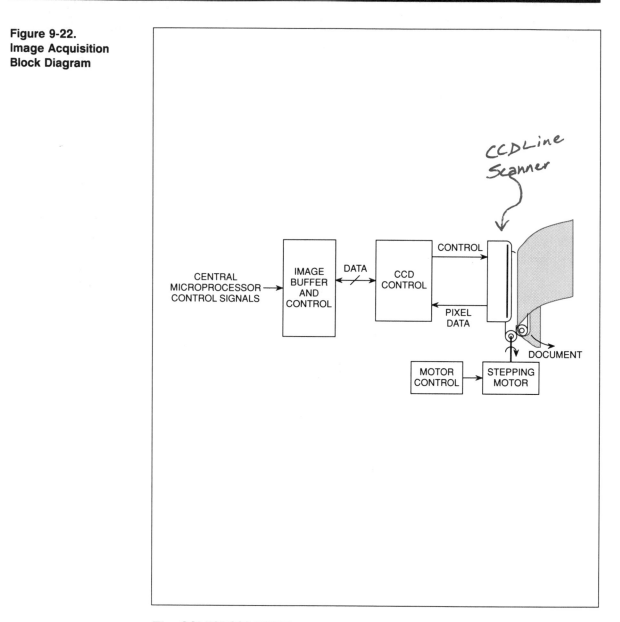

The SSI 73D2291/2292

The heart of all computer-fax applications is a fax modem which converts digital file information into analog telephone signals, and vice versa. Silicon Systems Incorporated has introduced the SSI 73D2291/2292 fax modem designed expressly for use in personal computer, portable computer, and laptop fax systems. A block diagram of the SSI 73D2291/ 2292 is shown in *Figure 9-23*.

**Figure 9-23.
SSI 73D2291/2292
Block Diagram.**
*(Courtesy Silicon
Systems, Inc.)*

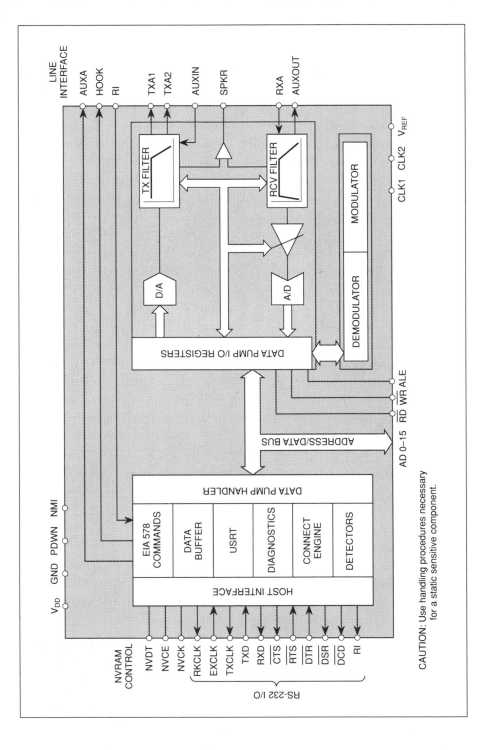

**Figure 9-24.
SSI 73D2291/2292 Fax
Modem.** *(Courtesy
Silicon Systems, Inc.)*

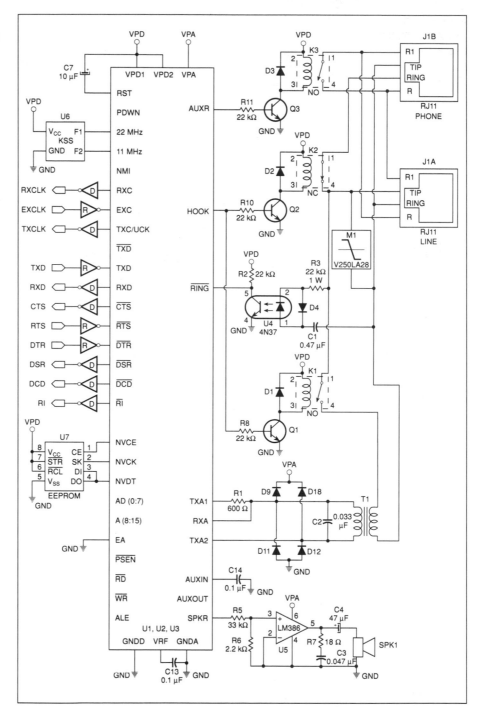

Communication is established to the host computer over a built-in RS-232 port. The SSI 73D2291/2292 will automatically set its own baud rate (autobaud) depending on the computer speed and the baud rate of the destination fax. It accepts high-level commands from the computer to control dialing, answering, speaker volume, and communication parameters such as synchronous/asynchronous data transfer, start and stop bit configuration, and pulse or DTMF dialing selection. A complete circuit to implement a computer-based fax is shown in *Figure 9-24*. All that is required is a UART for serial-to-parallel translation, a telephone line, and a software program to perform fax compression/decompression and manipulate the screen display or printout. Nonvolatile memory can be added to supply custom information or ID codes to the fax.

The SSI 73D2291/2292 frames its data using the High Level Data Link Control (HDLC) protocol. It also supports error checking and correction by adding a Frame Check Sum (FCS) to the data stream. Fax operating sequences are similar to (but are more complicated than) those required for conventional modems since fax operation needs format conversion and file compression steps in addition to normal data transfer.

WHAT HAVE WE LEARNED?

1. Modems change digital signals into signals that can be handled over standard telephone lines.
2. Digital data are formatted in particular serial sequences in order to transmit them asynchronously or synchronously.
3. When transmitting digital data over telephone lines, there is an originating modem on one end and an answering modem on the other.
4. Modems may have the capability for full-duplex, half-duplex, or simplex communication.
5. A modem modulates a carrier to transmit digital data.
6. A modem demodulates a carrier to convert transmitted data to the original digital data.
7. EIA standards for the U.S. and CCITT standards for Europe set the signal levels and pin connections for interfacing modems to data terminal equipment.
8. Protocols are rules for digital communications that govern how data terminal equipment is programmed to accomplish the communications.
9. Facsimile machines use modems as an integral part of their operation.

Quiz for Chapter 9

1. Modems are required to connect computers to telephone lines because:
   a. the telephone network bandwidth is too high.
   b. the telephone network will not pass direct current.
   c. telephone company rules require them.
   d. none of the above.

2. What kind of modulation is used in modems?
   a. phase modulation.
   b. frequency modulation.
   c. amplitude modulation.
   d. pulse modulation.
   e. all of the above.
   f. all except d.

3. The most common technique for binary data transmission is:
   a. synchronous transmission.
   b. bisynchronous transmission.
   c. asynchronous transmission.
   d. plesiochronous transmission.
   e. none of the above.

4. Asynchronous data transmission requires a clock:
   a. at the transmitter end.
   b. at the receiver end.
   c. at both ends.
   d. at neither end.
   e. all of the above.

5. The data transmission code most widely used in the U.S. is:
   a. Fielddata.
   b. Baudot.
   c. EBCDIC.
   d. ASCII.
   e. CCITT No. 5.

6. 103-type and 202-type modems use what modulation scheme?
   a. AFSK
   b. PSK
   c. MFM
   d. SSB-AM
   e. TDMA

7. The difference between asynchronous and synchronous transmission is:
   a. the way the synchronization is provided.
   b. the way the beginning of a block of data is detected.
   c. the way the beginning of a character is determined.
   d. all of the above.
   e. none of the above.

8. The parameter that most affects transmission of high-speed modem data is:
   a. phase distortion.
   b. amplitude distortion.
   c. frequency shift.
   d. impulse noise.

9. Error control of data transmission is done by:
   a. retransmission.
   b. adding redundancy.
   c. parity.
   d. cyclic redundancy checks.
   e. all of the above.

10. Protocols may be:
   a. bit oriented.
   b. byte oriented.
   c. character oriented.
   d. any of the above.
   e. none of the above.

Wireless Telephones

ABOUT THIS CHAPTER

This chapter focuses attention on telephones that perform most or all of the functions of the conventional telephone, but are connected with a radio link rather than wired directly. Although the term "wireless" may seem a bit old-fashioned, it is about the only general name for such telephones, since terms like "cordless telephone," "cellular telephones," "mobile telephone," and even "radiotelephone" have come to mean specific types of telephones without wires.

CORDLESS TELEPHONES

The base and portable units of cordless telephones are linked by a low power FM transmitter/receiver system.

The first type of wireless telephone to be discussed is the "cordless" which is used as an extension telephone in homes and businesses. *Figure 10-1* shows that the cordless telephone consists of two parts: a base unit and a portable unit. The connecting wires of a conventional telephone between the portable unit and the base unit are replaced with low-power radio transmissions. The radio link is completed by the transmission of a carrier that is frequency modulated (FM) with the information to be transmitted. The carrier and modulation principles are the same as described for the modem in Chapter 9; the only difference is that the frequency of the carrier is much higher for the cordless telephone.

The cordless telephone is one of two types. For one type, the portable unit has no keypad, and only the voice communications are radio linked. The keypad is on the base unit. The other type has the keypad on the portable unit and all the normal functions of a telephone are radio linked to the base unit.

Cordless telephones are available with the same features as electronic telephones that use cords. However, the cordless type must be plugged into a power source rather than operating from the line.

The telephones are electronic telephones as have been described in this book with the radio frequency transmission and modulating and demodulating electronic circuits added for the radio link. They have pulse generators and/or DTMF generators for dialing, electronic single- or dual-frequency ringers, and electronic speech circuits. Some have special features that allow the telephone to be used as an intercom. Some have speech and/or ringer volume controls and some have special security features to prevent unauthorized use. Most have the redial feature and some have a memory to store several numbers. The cord type electronic telephone usually obtains the operating power directly from the telephone line as has been described earlier; however, the cordless telephone, because of its greater power requirements, usually obtains operating

**Figure 10-1.
Cordless Telephone**

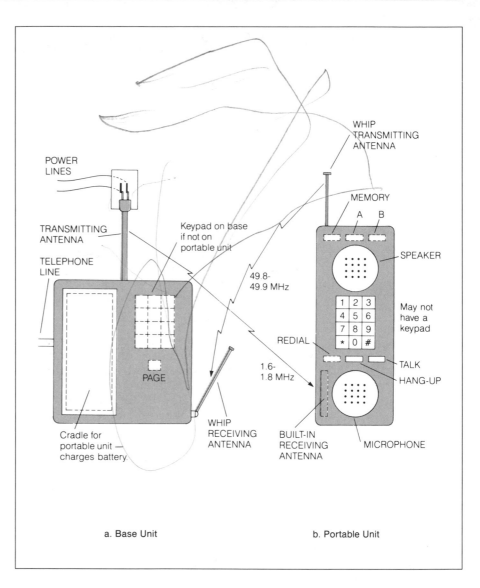

POWER
LINES

WHIP
TRANSMITTING
ANTENNA

TRANSMITTING
ANTENNA

Keypad on base
if not on
portable unit

MEMORY

A   B

TELEPHONE
LINE

49.8-
49.9 MHz

SPEAKER

1	2	3
4	5	6
7	8	9
*	0	#

May not
have a
keypad

REDIAL

1.6-
1.8 MHz

TALK

HANG-UP

PAGE

WHIP
RECEIVING
ANTENNA

BUILT-IN
RECEIVING
ANTENNA

MICROPHONE

Cradle for
portable unit —
charges battery.

a. Base Unit

b. Portable Unit

The base unit transmits on a frequency of 1.6 to 1.8 MHz. It uses the ac power line that supplies the power for the base station electronics as a transmitting antenna.

power from the household ac outlet. This is a minor disadvantage since the cordless telephone won't operate if a utility power failure occurs.

Base Unit

As shown in *Figure 10-1*, the base unit connects directly to the telephone line to complete the local loop to the central office. A two-wire to four-wire hybrid arrangement couples the local loop to the separate transmit and receive sections in the base unit. The base unit transmits on a carrier frequency in the range from 1.6 to 1.8 MHz and the household electrical wiring is used as the transmitting antenna for the base unit. The nominal 1.7 MHz frequency modulated signal is fed from the base unit transmitter to the ac line through capacitors which block the line current from the base unit transmitter while passing the 1.7 MHz output to the line. This use of the house wiring as an antenna is not unique to cordless telephones; it also is used for wireless intercoms. This method provides good reception within and near the house, as well as outside near power lines that are on the same side of the utility company's distribution transformer as the house circuit. This may include a neighbor's house wiring; thus, the potential exists for interference if that person also has a cordless telephone; more about this later.

Portable Unit

The portable unit receives the locally radiated RF signal from the base unit's antenna (the house wiring) with its built-in loopstick antenna, much like a portable radio.

An internal loopstick antenna (like that used in standard radio receivers) in the portable unit receives the nominal 1.7 MHz transmission from the base unit over a range from 50 to 1,000 feet. The range depends not only on the manufacturer's design, but also on such things as whether the house wiring is enclosed in metal conduit and whether foil-backed insulation is used in the walls. The ringing or voice signal is recovered by demodulation and drives the speaker in the portable unit. The portable unit is powered by a battery which is recharged when placed in a receptacle in the base unit.

The portable unit is kept in standby mode until a call is received. When a call is received, the portable unit transmits a 49.8 to 49.9 MHz signal to the base unit, which applies an off-hook signal to the local loop.

The portable unit is usually in a standby mode which corresponds to the on-hook condition of a telephone set. When the ringer sounds, the user operates a talk switch which turns on the transmitter in the portable unit. This transmitter transmits on a frequency in the range of 49.8 to 49.9 MHz and outputs the signal on the whip antenna. (Since the whip is used only for transmit, it can be collapsed out of the way when the portable unit is on standby. If the portable unit is only for voice transmission, it may have an internal antenna and its range is shorter.) A similar whip antenna on the base unit receives the FM signal from the portable unit, demodulates it, and applies the off-hook signal to the telephone local loop.

Dialing transmits modulation tones to the base unit, which sends either tones or pulses over the telephone lines. After the two parties have been successfully connected, the transmitter and receiver operate simultaneously.

When the user dials the number for outgoing calls, the dial pulses produce tones which modulate the carrier for transmission to the base unit. The base unit recovers the tones by demodulation. If DTMF service is used, the tones are sent on the telephone line. If pulse service is used, the tones are converted to pulses and the telephone line is pulsed. When the connections between calling and called parties are established, both transmitters and receivers operate at the same time to permit two-way conversation.

Frequencies

Although a few cordless telephones use the same frequency for transmission in both directions, most use two different frequencies in the ranges given above. Unless multiplexing is used, the use of a single frequency provides only half-duplex (one-way-at-a-time) transmission while the use of two frequencies allows full-duplex (simultaneous two-way) transmission just like a wired telephone. A choice of several frequencies is available so that neighbors can use different frequencies to prevent interference and eavesdropping. Also, signaling is done by guard tones in sequences which are selectable and unlikely to be duplicated in the neighborhood.

MOBILE TELEPHONES

Mobile telephones may be thought of as cordless phones with elaborate portable and base units. High-power transmitters and elevated antennas that provide the radio carrier link over an area within 20 to 30 miles from the base station antenna, as well as the multiplexing, detecting, sorting, and selecting features required to simultaneously service 60 subscribers per base station, are the major differences between cordless telephones and mobile telephones.

Base Unit

The mobile telephone base unit can operate on many channels simultaneously and can easily cover the average city with a power of several hundred watts.

Figure 10-2 shows a mobile telephone system. The base station can transmit and receive on several different frequencies simultaneously to provide several individual channels for use at the same time. The radio base station transmitter output power is typically 200–250 watts and the radiated power can be as high as 500 watts if the transmitting antenna gain is included. It covers a circular area of up to 30 miles in radius for clear reliable communications, but transmitters with the same frequency are not spaced closer than about 60 to 100 miles because of the noise interference levels.

The base unit receiver contains the necessary electronics to present its control terminal with a good audio signal. The control terminal interfaces the voice and control signals to the standard telephone circuits.

The receiver contains filters, high-gain amplifiers, and demodulators to provide a usable voice signal to the telephone line. The control terminal contains the necessary detector and timing and logic circuits to control the transmission link between the base unit and the mobile units. As a result, telephone calls are coupled to and from the standard telephone system just like calls that are carried completely over wired facilities. The control terminal has the necessary interface circuits so that a call initiated at a mobile unit is interconnected through the national or international telephone system to the called party just as any other telephone call.

The national and international telephone system facilities are owned by the respective telephone companies. The base units and mobile units may be owned by the telephone company or by a separate company called a radio common carrier (RCC). When the mobile system is run by an RCC, the RCC is charged by the telephone company for the use of the standard telephone system just like any other customer. This cost is then included in the charge by the RCC to the eventual user of the mobile units.

**Figure 10-2.
Mobile Telephone
System**

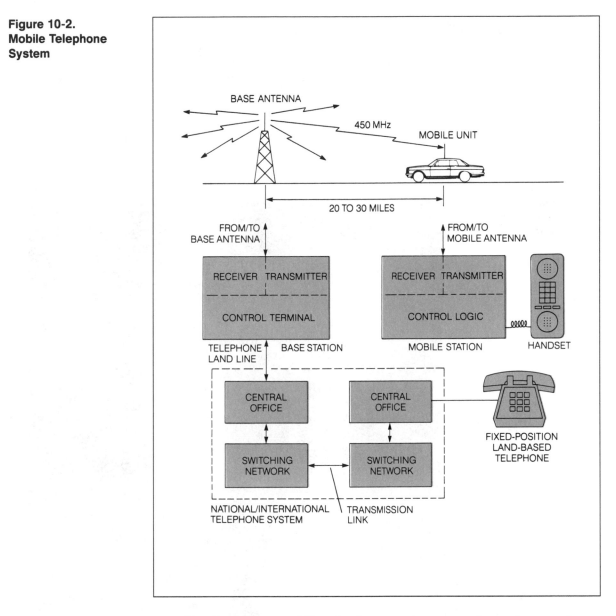

To subscribe to mobile telephone service, a user has only to apply
and be accepted by the RCC or the telephone company operating the
system. When the application is accepted, the user can lease or purchase
the mobile equipment.

Mobile Unit

The mobile unit contains a receiver, a transmitter, control logic, control unit, and antennas. For the user, it operates pretty much like an ordinary electronic telephone.

The mobile unit in the user's vehicle consists of a receiver containing amplifiers, a mixer, and a demodulator; a transmitter containing a modulator, carrier oscillators, and amplifiers; the necessary control logic; a control unit with microphone, speaker, keypad and switches; antennas and the interconnecting cables. The control unit performs all of the functions associated with normal telephone use. A modern control head with automatic functions is illustrated in *Figure 10-3*.

Figure 10-3.
Control Head for
Mobile Telephone
(Courtesy of Motorola, Inc.)

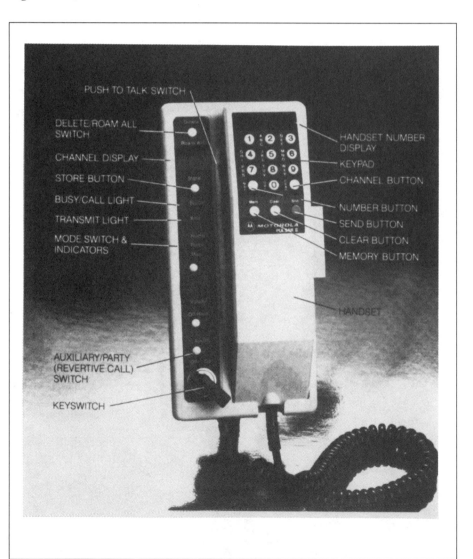

The mobile telephone user with automatic control places and receives calls in the same manner as with an ordinary telephone. When the handset is lifted to place a call, the radio unit automatically selects an available channel. If no channel is available, the busy light comes on. If a channel is found, the user hears the normal dial tone from the telephone system, and can then dial the number and proceed as if the telephone were direct wired. An incoming call to the mobile unit is signaled by a ringing tone and is answered simply by lifting the handset and talking. Thus, the automatic mobile telephone is as easily used as a home telephone. The mobile telephone combines the mobility of the radio link and the world-wide switched network of the existing telephone system to provide a communication link to any other telephone in the world.

Home Area and Roaming

If a subscriber goes outside the range of his base station, his mobile telephone can only be reached through another similar adjacent mobile base station system, provided advanced arrangements have been made.

As previously stated, the mobile system is designed for optimum use within a 20–30 mile radius of the base antenna. This is called the subscriber's home area and a subscriber usually will remain in the home area. However, if the subscriber travels out of the home area into another area, the subscriber is referred to as a roamer and a different mode of operation applies.

Each mobile telephone has a unique telephone number which includes the home area's base station identification. When someone calls the mobile unit, the calling party is connected first to the transmitter serving the subscriber's home area. As long as the subscriber is within radio range of that system, all is well; otherwise, the base station won't get an answer from the mobile unit and the caller will get a no-answer signal. If the subscriber roams outside the home area, he/she can still be reached if a similar mobile telephone system exists in that area, provided proper advance arrangements have been made.

Calls to roamers are usually placed by calling a special number for the mobile service operator who knows the roamer's location. The operator manually patches the call through to the base station serving the area of the roamer's location. Some systems cannot handle roamers due to overload of their channels, and some systems do not allow roamers.

Detailed Operation

For wireless operation, tones are used for those signaling functions otherwise performed by voltage and current in hard-wired systems.

Different signaling techniques must be used in a mobile telephone system than in a wired facility. Since there are no wires connecting the telephone to the network, both speech and signaling must be transmitted via radio. This is accomplished through the use of special tones rather than applying a voltage level or detecting a current. The tones are selected so as not to be mistaken for other signaling tones, such as DTMF. The proper tone transmitted to the mobile unit will, for example, ring the mobile telephone to indicate an incoming call just as with a standard telephone. A different tone is used to indicate off-hook, busy, etc.

The Improved Mobile Telephone System (IMTS) uses in-band signaling tones from 1,300 Hz to 2,200 Hz. The older Mobile Telephone

System (MTS) had in-band signaling tones in the 600 Hz to 1,500 Hz range. Some systems use 2,805 Hz in manual operation.

Incoming Call

To gain a better understanding of the system operation, let's trace an incoming call from a wire facility subscriber through the base unit to a mobile unit. The base station controls all activity on all channels and can transmit on any idle channel. Regardless of how many channels are idle, it selects only one and places a 2,000-Hz idle tone on it as shown in *Figure 10-4.* All on-hook mobile units that are turned on automatically search for the idle tone and lock on the idle channel because this is the channel over which the next call in either direction will be completed. After locking onto the idle channel, all on-hook mobile units "listen" for their number on that channel. When an idle channel becomes busy for a call in either direction, the base station control terminal selects another unused channel and marks it with the idle tone. All on-hook mobile units then move to the new idle channel. This process is repeated each time a new call is initiated as long as unused channels are available.

After the person calling the mobile subscriber dials the mobile unit's telephone number, the call is processed through the switched telephone network as in a normal landline call. The sequence in *Figure 10-4* is as follows: When the call reaches the control terminal, the terminal seizes the idle channel, and indicates seizure by removing the idle tone from that channel and applying the 1,800-Hz seize tone. The seize tone prevents other mobile units from seizing the channel to originate a call. The control terminal then out-pulses the mobile unit's number over the base station transmitter at ten pulses per second, with the idle tone representing a mark (which corresponds to the make interval in dc pulsing) and the seize tone representing a space (corresponding to the break interval).

Each on-hook mobile unit receiving the number transmission compares the received number to its unit number. As soon as a digit mismatch is detected, the mobile unit abandons that channel and searches for the new idle channel. Thus, upon completion of the number transmission, all mobile units except the one called will have abandoned the seized channel and will be monitoring the new idle channel.

When the mobile unit receives its correct seven-digit address, the mobile supervisory unit turns on the mobile transmitter and sends the acknowledgement signal, using the 2,150-Hz guard tone, back to the control terminal. If this acknowledgement is not received by the control terminal within three seconds after out-pulsing the address, the seize tone is removed and the call abandoned. However, upon receipt of the mobile acknowledgement signal, the terminal sends standard repetitive ringing at a cycle of two seconds on, four seconds off, using idle and seize tones as before. If the mobile does not answer within 45 seconds, ringing is discontinued and the call abandoned.

The base station selects one idle channel and modulates it with a 2-kHz tone. This becomes a marked idle channel. This channel has now been reserved for the next land-originating telephone call.

When the land-originating call reaches the control terminal, the idle tone is replaced by a 1.8-kHz seize tone which seizes the reserved channel. The control terminal then sends out via the transmitter the mobile unit's number.

Only the one mobile unit with a number match remains locked on that channel.

When it receives its unique number, the mobile transmitter automatically broadcasts a 2.150-kHz acknowledgement. Ringing signals are then broadcast to the mobile unit.

**Figure 10-4.
Base to Mobile
Sequence**

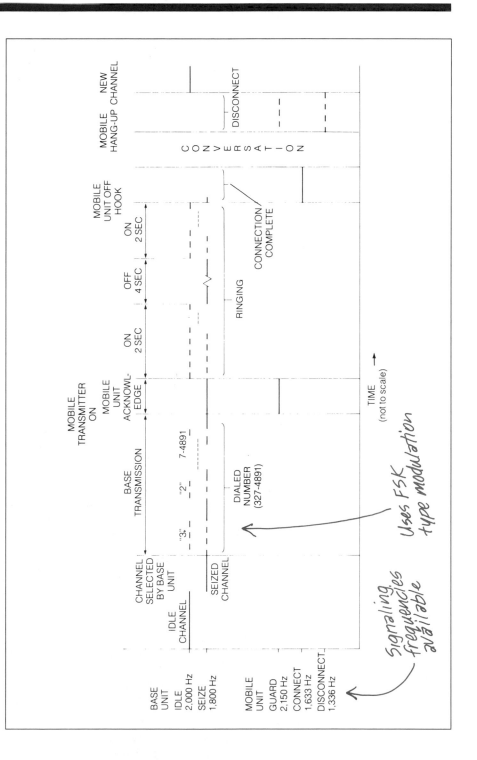

If the mobile unit goes off-hook to answer, another tone frequency burst is sent, allowing voice communications to commence. Upon hang-up (on-hook), the mobile unit sends a disconnect signal by alternating disconnect tone and tone guard.

When a call is originated from the field, the mobile unit finds a marked idle channel and broadcasts an acknowledgement to the base by sending its identification. The mobile unit then completes a call in the usual manner by receiving a dial tone, dialing the number, and waiting for the called party to answer.

When the mobile subscriber goes off-hook to answer, the mobile supervisory unit sends a burst of connect tone (1,633 Hz) as an answer signal. Upon receipt of the answer signal, the control terminal stops the ringing and establishes a talking path between the calling circuit and the radio channel. When the subscriber hangs up at the end of the call, the mobile supervisory unit sends a disconnect signal—alternating the disconnect tone (1,336 Hz) and the guard tone. The mobile supervisory unit then turns off the mobile transmitter and begins searching for the marked idle channel.

Outgoing Call

The sequence for a call originated by a mobile subscriber is illustrated by *Figure 10-5*. When the subscriber goes off-hook to place the call, the mobile unit must be locked on the marked-idle channel. If not, the handset will be inoperative and the busy lamp on the control unit will light, indicating to the subscriber that no channel is available. If the mobile unit is locked on the marked-idle channel, the mobile supervisory unit will turn on the mobile transmitter to initiate the acknowledgement or handshake sequence. The identification section of *Figure 10-5* is where the mobile unit transmits its own number so the control terminal can identify it as a subscriber and can charge the call to the number. The pulses of guard tone mixed in with the number pulses are for parity checking. The remaining functions of *Figure 10-5* are similar to those of *Figure 10-4*.

CELLULAR MOBILE TELEPHONE SERVICE

Mobile telephone service has always been a scarce luxury. Subscribers pay from ten to twenty times more for mobile service than for residential telephone service, yet most urban telephone carriers and RCCs have long waiting lists for mobile telephones. In Chicago, for example, only 2,000 mobile users can be accommodated, yet at least ten times that many desire service at present rates. The reason is that there simply are not enough channels to handle the demand, and the few dozen available are spread over several bands and divided among different types of carriers. The solution is not simply to assign new frequencies and build more transmitters because the spectrum space for new frequencies is simply not available; besides, this would not eliminate the restrictions on roamers. Clearly, an entirely new approach to mobile telephony was needed. The cellular concept, also called the Advanced Mobile Phone Service (AMPS)[1] is a method to provide high quality mobile service for more subscribers at an affordable cost and to provide more freedom for roamers.

---

[1]*Advanced Mobile Phone Service (AMPS) is a registered service mark of AT&T Co.*

**Figure 10-5.
Mobile to Base
Sequence**

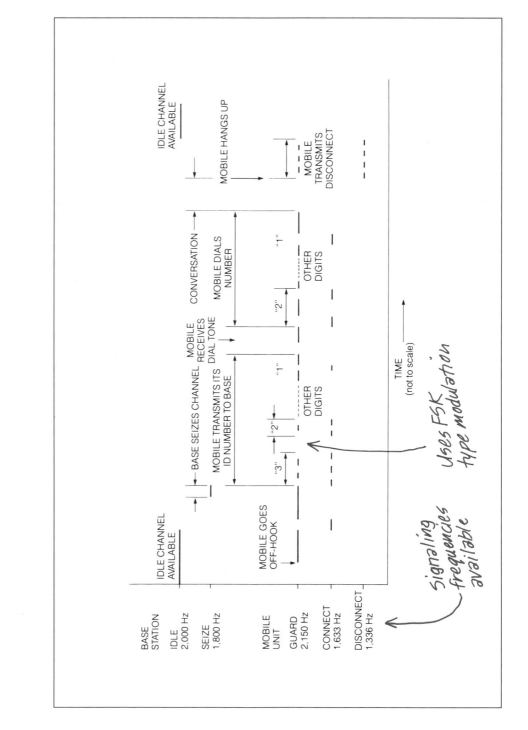

Basic Concept

The basic concept of the AMPS cellular system is to reduce the area covered by the transmitter by reducing the power of transmission. In this way, concentrated areas of population can have more transmitting stations, and thus more channels, because each transmitter handles a given number of conversations. In addition, because transmitters cover less area, the same frequency can be reused in a common geographical area.

System Structure

By dividing a city into many cells, each serviced by a low power transceiver base station, the number of available channels over the city is increased enormously.

The basic system arrangement is shown in *Figure 10-6*. The service area is divided into regions called cells, each of which has equipment to switch, transmit, and receive calls to and from any mobile unit located in the cell. Each cell transmitter and receiver operates on a given channel. Each channel is used for many simultaneous conversations in cells which are not adjacent to one another, but are far enough apart to avoid excessive interference. Thus, a system with a relatively small number of subscribers can use large cells, and as demand grows, the cells are divided into smaller ones.

The Cell Site

Cell sites form the radio link between individual cellular telephones and the telephone system. Each cell station is equipped with a transmitter and receiver coupled to an array of antennas as shown in *Figure 10-7*. Telephone network switching electronics, as well as support and diagnostic electronics are included at each site. Cells are located where they will operate most effectively in the radio environment. In urban areas, they may be found atop tall buildings. Suburban or rural cell sites may be located on large hills or mountains—wherever the best radio coverage can be obtained.

A typical cell site is designed to handle as many as 45 two-way conversations. Since each conversation requires two frequencies for full-duplex operation, each cell will use up to 90 of the 666 available frequencies. Adjacent cells will use other frequency sets as suggested in *Figure 10-8*. As an example, suppose the center of Area 1 uses frequencies 1 through 90. No adjacent cells can use those same frequencies because of possible interference, so adjacent cells will be assigned other frequency sets. The top cell (1–90) of Area 2 is a nonadjacent cell. It is far enough away to prevent any interference, so it may reuse the same frequencies as the center cell of Area 1.

**Figure 10-6.
Cellular Network**

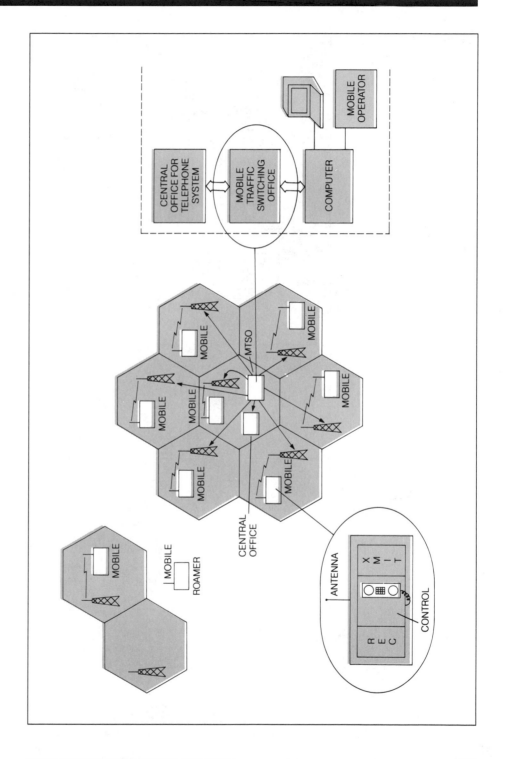

**Figure 10-7.
A Typical Cell Site.**

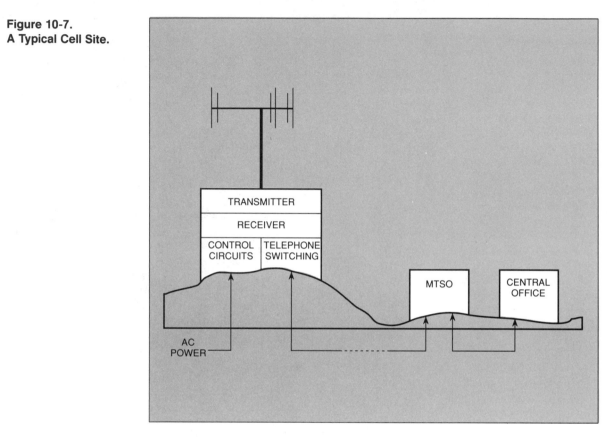

**Figure 10-8.
Example of Cellular
Frequency Use.**

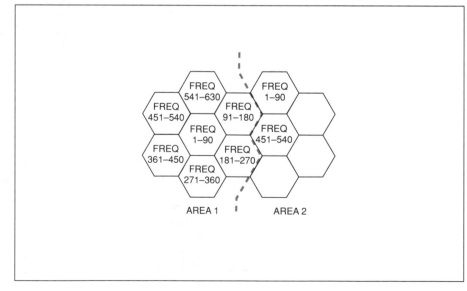

The MTSO

A central Mobile Tele-communication Switching Office (MTSO) performs all of the functions of a normal switching office and also controls each of the cell transceiver functions.

The cell sites are interconnected and controlled by a central Mobile Telecommunications Switching Office (MTSO), which is basically a telephone switching office as far as hardware is concerned, but as shown in *Figure 10-6*, it uses a substantial amount of additional digital equipment programmed for cellular control. It not only connects the system to the telephone network, but also records call information for billing purposes. The MTSO is linked to the cell sites by a group of voice trunks for conversations, together with one or more data links for signaling and control. The MTSO controls not only the cell sites via radio commands, but also many functions of the mobile units.

Mobile Units

The mobile units consist of a control unit, a transceiver, and appropriate antennas. The transceiver contains circuits that can tune to any of the 666 FM channels allotted by the FCC from 826 to 845 MHz and 870 to 890 MHz in the cellular range. Each cell site has at least one setup channel dedicated for signaling between the cell and its mobile units. The remaining channels are used for conversation. Each channel is 30 kHz wide, and two channels are required for full duplex operation.

Each mobile unit is assigned a 10-digit number, identical in form to any other telephone number. Callers to the mobile unit will dial the local or long-distance number for the desired mobile unit. The mobile user will dial seven or ten digits with a 0 or a 1 prefix, where applicable, as if calling from a fixed telephone.

Whenever a mobile unit is turned on but not in use, the mobile control unit monitors the data being transmitted on a setup channel selected from among the several standard setup frequencies on the basis of signal strength. If signal strength becomes marginal as the mobile unit approaches a cell boundary, the mobile control finds a setup channel with a stronger signal.

Using a Cellular Telephone

The actual process of using a cellular telephone is very similar to that of a regular home telephone.

Placing a call begins with a request for service. This is done by taking the cellular telephone off-hook. A cellular transmitter seizes an available set-up channel and sends a service request to the serving cell site. The cell site assigns a voice channel set over which dialing and voice signals will take place. The cellular control circuit automatically switches to this "conversation" channel and a voice link is now established. Unlike regular telephones, dial tone signals are not transmitted over cellular channels. A visual display is usually employed instead to indicate that a channel is ready. Dialing can now take place normally. A user may either dial the desired number while on-line, or recall a presaved number from the telephone memory. The cell site interprets incoming dial tones and sends the digits along to the MTSO which will connect the cellular telephone to its destination. Ringback or busy tones will be sent from the

MTSO through the cell site to the cellular telephone. If the destination telephone goes off-hook, a conversation will take place normally.

Receiving a call reverses this process. A central office will acknowledge an off-hook subscriber telephone with a dial tone. The caller will dial the number of the desired cellular telephone. Central office circuits will connect to the appropriate MTSO which will order each cell to transmit the necessary code number for the cellular unit on a set-up channel. When the desired cellular telephone recognizes its unique code, it will automatically seize the set-up channel of the nearest cell and acknowledge that it is ready. Cell control circuits select an available voice channel set and order the cellular unit to switch over. After a voice channel link is established, the cell will signal the desired cellular telephone of an incoming call by ringing the telephone. If the cellular telephone goes off-hook, a normal conversation can take place.

Whether placing or receiving a call, cell circuitry monitors the strength of the cellular signals every few seconds. Signal strength information is sent to the MTSO which evaluates the nearest available cell site. As the cellular signal falls off in one cell, the MTSO orders another cell to establish a new voice channel and take over the call. This transfer of control and voice channel is known as "handoff." In actual operation, handoff takes place so fast that the cellular user does not even know that it has happened.

Roamers

The mobile unit automatically hands off the call to a cell and a channel that provides the optimum communications quality. The MTSO computer continuously analyzes signal quality and makes the appropriate changes without any interruption in service.

The system is designed to make handling of roamers automatic; indeed, this is the principal goal of the cellular approach. Locating and handoff are concepts that come directly from the use of small cells. "Locating" in this sense is not the determination of precise geographic location—although that is obviously a factor; rather, it is the process of determining whether a moving active user should continue to be served by his current channel and transmitter, or "handed off" to either another channel, cell, or both. The decision is made automatically by a computer, based on signal quality and potential interference, and involves sampling the signal from the mobile unit.

With the cellular system, a subscriber could make a call from his car while driving in the countryside toward a city, continue through the city's downtown, and not hang up until well beyond the city on the other side. During the entire time, the transmission would be clear. More importantly, the switching of transmitters and frequencies during the conversation would be entirely automatic, with no interruptions and no action required by the user or an operator.

Many cellular services have entered into mutual service agreements with other regional cellular providers. This allows service when roaming outside the primary service area. Wherever there is a system to serve it, a roaming unit will be able to obtain completely automatic service; however, a call from a land telephone to a mobile unit which has roamed to another metropolitan area presents additional problems. While it would be technically possible for the system to determine automatically where the mobile unit is, and to connect it automatically to the land party, there are two reasons for not doing so. First, the caller will expect to pay only a local charge if a local number is dialed. Second, the mobile user may not want to be identified to be at a particular location automatically by the system without an approval. Therefore, the system will complete the connection only if the extra charge is agreed to, and when possible to do so without unauthorized disclosure of the service area to which the mobile unit has roamed.

Unique Features

Multiple frequency re-use is possible because of the lower transmitter power radiated in each cell, and by not using the same frequency in adjacent cells.

There are two essential elements of the cellular concept which are unique; frequency reuse and cell splitting.

Frequency reuse means using the same frequency or channel simultaneously for different conversations, in the same general geographic area. The idea of having more than one transmission on a given frequency is not new; it is done in virtually all radio services. What is unique to cellular is the closeness of the users; two users of the same frequency may be only a few dozen miles apart, rather than hundreds of miles. This is done by using relatively low-power transmitters on multiple sites, rather than a single high-power transmitter. Each transmitter covers only its own cell, and cells sufficiently far apart may be using the same frequency.

The cellular system can be expanded because cell splitting may occur as demand increases.

Cell splitting is based on the notion that cell sizes are not fixed, and may vary in the same area or over time. The principle is shown in *Figure 10-9*. Initially, all the cells in an area may be relatively large as shown in *Figure 10-9a*. When the average number of users in some cells becomes too large to be handled with proper service quality, the overloaded cells are split into smaller cells by adding more transmitters, as shown in *Figure 10-9b*. The same MTSO can continue to serve all of the cell sites, but expansion of its computer and switching facilities probably will be required.

Cellular Difficulties

By their very nature, cellular telephones have to be rugged and reliable. They are subjected to physical abuses such as drops and bumps, as well as extremes of heat and cold. Although the design and components used in many cellular telephones will provide long, reliable service, there are some situations that can impair cellular performance. These problems are usually related to the radio link between the telephone and the cell site.

Dropouts

Radio signals in the 800 to 900 MHz range tend to move in straight lines. They can be weakened by water in the air and reflected by buildings and naturally occurring objects such as hills or mountains. As a result, the signal strength may fall off enough in some instances to cause a momentary loss of the transmitted or received signal (or both). Cell site

**Figure 10-9.
Cell Splitting**
*(Reprinted with
permission from* Bell
System Technical
Journal, *Copyright ©
1979, American
Telephone and
Telegraph Co.)*

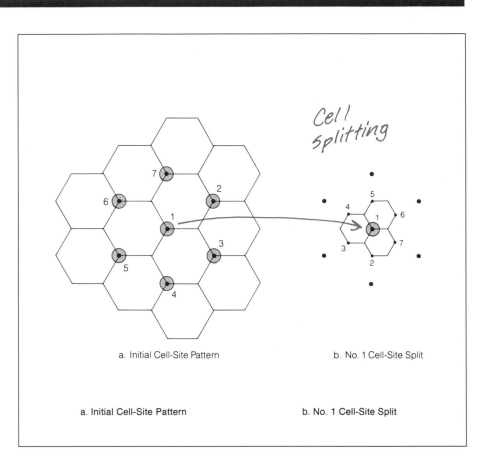

a. Initial Cell-Site Pattern   b. No. 1 Cell-Site Split

control circuits are designed to disregard dropouts so they will not be
interpreted as a hang-up condition. Under most circumstances, dropouts
cause little more than an annoyance.

Dropouts can also happen in fringe areas of cellular service where
there are no more cell sites to handoff the conversation to. In these
situations, the user will experience gradual weakening of the signal.
Dropouts will start as brief interruptions, but quickly worsen until a
disconnection occurs. The best attempt at avoiding dropoffs is to try
different travel paths if possible.

Dead Zones

In principle, dead zones are caused by the same general conditions as
dropouts, but on a much larger scale. Dead zones are common in hilly or
mountainous regions. Large buildings in urban areas can block enough
signal to create man-made dead zones. Radio signals can reflect off
multiple buildings or other objects and interfere with each other when
reaching the cell site. This phenomenon is usually known as "multipath"
interference. Since dead zones cause signal losses for much longer periods

of time, cell site controls will interpret this interference as a hang-up and terminate the conversation.

Other Cellular Systems

There are differences between cellular systems in the United States and Europe. European cellular systems operate from 890 to 989 MHz and each channel occupies and bandwidth of 25 kHz. The United Kingdom has been using two systems since 1985: the "Cellnet" system developed by British Telecom and Securicor, and the "Vodaphone" system created by the Racal Corporation. Both of these systems are designed to conform to the Total Access Communication System (TACS) used in Europe. TACS is based on AT&T's AMPS system used in the U.S.

Canada is developing a system to operate in the 800-MHz range. Cellular facilities are also being installed in the Middle East and East Asia.

WHAT HAVE WE LEARNED?

1. Wireless telephones consist of a base unit and a mobile unit that are connected by a radio link instead of wires.
2. Wireless telephones may be cordless phones for the home or small business, or mobile radio telephones used in commerce and industry.
3. Most cordless telephones use frequency modulation to carry signaling and voice communications.
4. Many mobile telephones have a high-power transmitter at the base station which covers an area with a radius of 20–30 miles.
5. Specific sequences and tone signaling must be followed to complete the radio link correctly between a mobile telephone base and mobile unit.
6. The cellular wireless telephone system was designed to allow many more subscribers to use wireless telephones.
7. A cellular system handles the switching of cell sites and signal channels automatically as a roamer moves through the grid of cells.

Quiz for Chapter 10

1. As used in this book, the term "wireless telephone" means:
  a. transoceanic telephone service via radio.
  b. a telephone where the wiring has been replaced by solid-state integrated circuits.
  c. any telephone device or system which uses a radio link to replace a wired link somewhere between the handset and the public switched network.

2. A mobile telephone is:
  a. another name for a cordless telephone.
  b. a special form of radio telephone that connects with the public switched telephone network through the telephone company.
  c. cordless extension telephone designed for use in automobiles.

3. Mobile telephone users originate and receive calls:
  a. by going through the regular telephone operator.
  b. automatically, provided they are in their home area and a channel is available.
  c. by using two different radio channels, one for transmission and one for reception.

4. Cellular mobile telephone service is:
  a. the concept of using many low power transmitters with computer control, rather than a few high-power transmitters.
  b. another name for IMTS.
  c. the concept of improving frequency utilization by restricting users to a single or home area called a cell.

5. A cordless telephone:
  a. uses fiber-optic techniques to replace the traditional cord.
  b. is a portable extension telephone using short-distance radio links between the handset and the base unit.
  c. uses radio links between the base unit and the telephone company lines.

6. Antennas for a cordless telephone:
  a. are usually placed on the roof, sometimes combined with the TV antenna.
  b. consist of two whips, one for transmission from the base, the other for transmission from the portable.
  c. consist of the building wiring, whips, and loopsticks.

7. A roamer is:
   a. a user of a cordless telephone who walks around the building while talking.
   b. a term that applies only to cellular systems.
   c. a mobile telephone user who is outside of the home area.

8. When a cellular mobile telephone moves from one cell to another while a call is in progress,
   a. the call must be terminated, then initiated from the new cell.
   b. the system automatically transfers the call to another transmitter, and possibly to another channel.
   c. the user must make arrangements with the telephone company to have calls forwarded.

9. An idle channel in the IMTS system is:
   a. the condition of each mobile unit's dedicated channel when it is not being used.
   b. one unused channel which all mobile units monitor for their address.
   c. a special channel dedicated for signaling only, which is often idle.

10. Cell splitting is:
    a. the process of dividing a large mobile telephone cell into smaller cells when usage requires.
    b. the idea of two mobile units using the same frequency at different locations in the same cell.
    c. a process similar to splitting hairs, often used by people who make up multiple-choice quizzes.

Glossary

A/D Converter: A circuit that converts signals from analog form to digital form.

Address: The number dialed by a calling party which identifies the party called. Also a location or destination in a computer program.

Aliasing: The occurrence of spurious frequencies in the output of a PCM system that were not present in the input—due to foldover of higher frequencies.

AM (Amplitude Modulation): A technique for sending information as patterns of amplitude variations of a carrier sinusoid.

Amplifier: An electronic device used to increase signal power or amplitude.

Analog: Information represented by continuous and smoothly varying signal amplitude or frequency over a certain range, such as in human speech or music.

Asynchronous: Refers to circuitry and operations without common timing (clock) signals.

Attenuation: The decrease in power that occurs when any signal is transmitted.

Audio Frequency: Frequencies detectable by the human ear, usually between 20 and 15,000 Hz.

Bandwidth: The range of signal frequencies that a circuit or network will respond to or pass.

Base Unit: The transmitter (antenna and equipment), in a fixed location, and usually having higher power than the mobile units.

Binary Code: A pattern of binary digits (0 and 1) used to represent information such as instructions or numbers.

Bipolar: Having both positive and negative polarity.

Bit: An acronym for binary digit; the smallest piece of binary information; a specification of one of two possible alternatives.

BORSCHT: An acronym for the functions that must be performed in the central office when digital voice transmission occurs: *B*attery, *O*vervoltage, *R*inging, *S*upervision, *C*oding, *H*ybrid, and *T*est.

Byte: A group of 8 bits treated as a unit. Often equivalent to one alphabetic or numeric character.

Cable: An assembly of one or more conductors insulated from each other and from the outside by a protective sheath.

Cell: In cellular mobile telephony, the geographic area served by one transmitter. Subscribers may move from cell to cell.

Central Office (CO): The switching equipment that provides local exchange telephone service for a given geographical area, designated by the first three digits (NNX or NXX) of the telephone number.

Channel: An electronic communications path, usually of 4,000 Hz (voice) bandwidth.

Circuit: An interconnected group of electronic devices, or the path connecting two or more communications terminals.

Common Battery: A system of supplying direct current for the telephone set from the central office.

Compander: An acronym for COMpressor-exPANDER, a circuit that compresses the dynamic range of an input signal, and expands it back to almost original form on the output.

Crossbar Switch: An electromechanical switching machine utilizing a relay mechanism with horizontal and vertical input lines (usually 10 by 20), using a contact matrix to connect any vertical to any horizontal.

Crosspoint: The element that actually performs the switching function in a telephone system. May be mechanical using metal contacts, or solid state using integrated circuits.

Crosstalk: Undesired voice-band energy transfer from one circuit to another (usually adjacent).

Current: The flow of electrical charge, measured in amperes.

Cut-Off Frequency: The frequency above which or below which signals are blocked by a circuit or network.

D/A Converter: A circuit that converts signals from digital form to analog form.

Data: In telephone systems, any information other than human speech.

Data Set: Telephone company term for modem.

Decibel (dB): A unit of measure of relative power or voltage, in terms of the ratio of two values. $dB = 10 \log P1/P2$, where P1 and P2 are the power levels in watts.

Decoder: Any device which modifies transmitted information to a form which can be understood by the receiver.

Demodulation: The process of extracting transmitted information from a carrier signal.

Demultiplexer: A circuit that distributes an input signal to a selected output line (with more than one output line available).

Digital: Information in discrete or quantized form; not continuous.

Distortion: Any difference between the transmitted and received waveforms of the same signal.

DTMF (Dual-Tone-Multi-Frequency): Use of two simultaneous voice-band tones for dialing.

Electromagnetic Spectrum: The entire available range of sinusoidal electrical signal frequencies.

Encoder: Any device which modifies information into the desired pattern or form for a specific method of transmission.

ESS (Electronic Switching System): A telephone switching machine using electronics, often combined with electromechanical crosspoints, and usually with a stored-program computer as the control element.

Exchange Area: The territory within which telephone service is provided without extra charge. Also called the Local Calling Area.

FCC (Federal Communications Commission): The U.S. government agency that regulates and monitors the domestic use of the electromagnetic spectrum for communications.

Fiber Optics: The process of transmitting infrared and visible light frequencies through a low-loss glass fiber with a transmitting laser or LED.

FM (Frequency Modulation): A technique for sending information as patterns of frequency variations of a carrier signal.

Frequency: The rate in hertz (cycles per second) at which a signal pattern is repeated.

FSK (Frequency-Shift Keying): A method of transmitting digital information that utilizes two tones; one representing a one level, the other a zero level.

Ground: An electrical connection to the earth or to a common conductor which is connected to the earth at some point.

Ground Start: A method of signaling between two machines where one machine grounds one side of the line and the other machine detects the presence of the ground.

Half-Duplex: A circuit that carries information in both directions, but only in one direction at a time.

Home Area: The geographic area in which a mobile telephone subscriber is normally located.

Hybrid: In telephony, a circuit that divides a single transmission channel into two, one for each direction; or conversely, combines two channels into one.

Instruction Code: Digital information that represents an instruction to be performed by a computer.

Integrated Circuit: A circuit whose connections and components are fabricated into one integrated structure on a certain material such as silicon.

Lineside: Refers to the portion of the central office that connects to the local loop.

Local Loop: The voice-band channel connecting the subscriber to the central office.

Loop Start: The usual method of signaling an off-hook or line seizure, where one end closes the loop and the resulting current flow is detected by the switch at the other end.

Loss: Attenuation of a signal from any cause.

Message Telephone Service (MTS): The official name for long distance or toll service.

Microwaves: All frequencies in the electromagnetic spectrum above one billion hertz (1 gigahertz).

Mobile Unit: The part of the mobile telephone system that is not fixed in its location; hence, is not attached to the telephone network by wires.

Modulation: The systematic changing of the properties of an electronic wave, using a second signal, to convey the information contained in the second signal.

Multiplexing: The division of a transmission facility into two or more channels.

Off-Hook: The condition that indicates the active state of a customer telephone circuit. The opposite condition is On-Hook.

Oscillator: An electronic device used to produce repeating signals of a given frequency and amplitude.

PABX or PBX: A private (automatic) branch telephone exchange system providing telephone switching in an office or building.

Parallel Data: The transfer of data simultaneously over two or more wires or transmission links.

Parity: A bit that indicates whether the number of "ones" in a bit string is odd or even.

PCM (Pulse Code Modulation): A communication systems technique of coding signals with binary codes to carry the information.

Period: The time between successive similar points of a repetitive signal.

Phase: The time or angle that a signal is delayed with respect to some reference position.

Portable Unit: The portion of a cordless telephone that is not electrically attached to the network by wires. Consists of a transmitter, receiver, and perhaps a keypad.

POTS: "Plain Old Telephone Service," single line residential rotary dial service.

Program: The sequence of instructions stored in the computer memory.

RCC (Radio Common Carrier): A company that provides mobile telephone service, but is not a telephone company.

Receiver: The person or device to which information is sent over a communication link.

Register: A series of identical circuits placed side-by-side that are able to store digital information.

Ring: The alerting signal to the subscriber or terminal equipment; the name for one conductor of a wire pair designated by R.

Roamer: A mobile telephone subscriber using the system outside of his/her home area.

Serial Data: The transfer of data over a single wire in a sequential pattern.

Sidetone: That portion of the talkers voice which is fed back to his/her receiver.

Simplex: A circuit which can carry information in only one direction; for example, broadcasting.

SLIC (Subscriber Line Interface Circuit): In digital transmission of voice, the circuit which performs some or all of the interface functions at the central office. See BORSCHT.

State: A condition of an electronic device, especially a computer which is maintained until an internal or external occurrence causes change.

Step-by-Step (SxS) System: An electromechanical telephone switching system in which the switches are controlled directly by digits dialed by the calling party.

Subscriber: The telephone customer.

Subscriber Loop: Another term for local loop.

Synchronous: Events that are controlled by or referred to a common clock.

TDM (Time Division Multiplexing): A communication system technique that separates information from channel inputs and places them on a carrier in specific positions in time.

Tip: One conductor of a wire pair, designed by T; usually the more positive of the two.

Toll Center: A major telephone distribution center that distributes calls from one major metropolitan area to another.

Transmission: Passing information, using electromagnetic energy, from one point to another.

Transmission Link: The path over which information flows from sender to receiver.

Transmitter: The person or device that is sending information over a communication link.

Trunk: A transmission channel connecting two switching machines.

Trunkside: The portion of the central office that connects to trunks going to other switching offices.

Voice-Grade Line: A local loop, or trunk, having a bandpass of approximately 300 to 3,000 Hz.

Voltage: A measure of the electrical force that causes current flow in a circuit.

Wideband Circuit: A transmission facility having a bandwidth greater than that of a voice-grade line.

Index

V

V.22 standard, 305, 307
V.22bis standard, 305
V.32 modem, 306
V.42 modems, 307
V.42bis modems, data compression, 307
Vodaphone system, 341
voice channel
   bandwidth, 14-15
   level, 15-18
   noise, 18
voice-frequency filters, 226-227
   Texas Instruments TMC2912C, 226-227
voice-grade line, 349
voltage, 349
   calculating for circuits, 109-110
   multitone ringer regulation, 130
   single-tone ringer regulation, 126
   transients, electronic dialing and ringing circuits, 108-110
volume, compensation in telephone set, 100-101

W

Watson, Thomas A., 1, 8, 63
waveform coders, 184-198
   coding, 188-198
      μ-law compander, 194
      A-law compander, 194-195
      companding, 192-194
      Delta Modulation, 195-197
      linear coders, 189-191
      source coders, 198
   pulse amplitude modulation (PAM), 186-187

pulse code modulation (PCM), 187
   quantization, 188
   sampling analog wave, 185-186
waveguide, 41
wideband circuit, 349
wire center, 11
wire medium, 39
wireless telephones, 323-341
   cellular mobile telephone service, 332, 334-341
   cordless telephones, 323-326
   mobile telephones, 326-333
word interleaving
   synchronization, 201-202
   time division multiplexing, 199-206

Z

zener diodes, 84
zero transmission level point (0 TPL), 17

Answers to Quizzes

Chapter 1

1. b
2. a
3. d
4. d
5. b
6. d
7. a
8. c
9. b
10. d
11. d
12. b
13. d

Chapter 2

1. b
2. a
3. b
4. b
5. d
6. c
7. a
8. d
9. b
10. d

Chapter 3

1. c
2. b
3. c
4. a
5. c
6. d
7. c
8. d
9. a
10. b

Chapter 4

1. c
2. d
3. a
4. c
5. a
6. d
7. b
8. d
9. c
10. b

Chapter 5

1. c
2. d
3. d
4. c
5. c
6. b
7. e
8. b
9. b
10. b

Chapter 6

1. b
2. d
3. d
4. b
5. c
6. c
7. d
8. d
9. b
10. a
11. c
12. b
13. b
14. a
15. b

Chapter 7

1. c
2. c
3. b
4. a
5. b
6. c
7. c
8. b
9. a
10. b

Chapter 8

1. e
2. d
3. c
4. e
5. a
6. a
7. d
8. a
9. b
10. d

Chapter 9

1. b
2. f
3. c
4. c
5. d
6. a
7. d
8. a
9. e
10. d

Chapter 10

1. c
2. b
3. b
4. a
5. b
6. c
7. c
8. b
9. b
10. a